# RILEM State-of-the-Art Reports

# RILEM STATE-OF-THE-ART REPORTS
Volume 47

RILEM, The International Union of Laboratories and Experts in Construction Materials, Systems and Structures, founded in 1947, is a non-governmental scientific association whose goal is to contribute to progress in the construction sciences, techniques and industries, essentially by means of the communication it fosters between research and practice. RILEM's focus is on construction materials and their use in building and civil engineering structures, covering all phases of the building process from manufacture to use and recycling of materials. More information on RILEM and its previous publications can be found on www.RILEM.net.

The RILEM State-of-the-Art Reports (STAR) are produced by the Technical Committees. They represent one of the most important outputs that RILEM generates – high level scientific and engineering reports that provide cutting edge knowledge in a given field. The work of the TCs is one of RILEM's key functions.

Members of a TC are experts in their field and give their time freely to share their expertise. As a result, the broader scientific community benefits greatly from RILEM's activities.

RILEM's stated objective is to disseminate this information as widely as possible to the scientific community. RILEM therefore considers the STAR reports of its TCs as of highest importance, and encourages their publication whenever possible.

The information in this and similar reports is mostly pre-normative in the sense that it provides the underlying scientific fundamentals on which standards and codes of practice are based. Without such a solid scientific basis, construction practice will be less than efficient or economical.

It is RILEM's hope that this information will be of wide use to the scientific community.

Indexed in SCOPUS, Google Scholar and SpringerLink.

Kefei Li · Junjie Zeng · Luping Tang ·
Mette Rica Geiker · Riccardo Maddalena
Editors

# Long-Term Performance of Structural Concretes in Marine Exposure Conditions

State-of-the-Art Report of the RILEM Technical Committee 289-DCM

Springer

*Editors*
Kefei Li
Department of Civil Engineering
Tsinghua University
Beijing, China

Luping Tang
Department of Architecture and Civil Engineering
Chalmers University of Technology
Gothenburg, Sweden

Riccardo Maddalena
School of Engineering
Cardiff University
Cardiff, UK

Junjie Zeng
CCCC Fourth Harbor Engineering Co. Ltd.
Guangzhou, China

Mette Rica Geiker
Department of Structural Engineering
NTNU
Trondheim, Norway

ISSN 2213-204X ISSN 2213-2031 (electronic)
RILEM State-of-the-Art Reports
ISBN 978-3-032-06958-0 ISBN 978-3-032-06959-7 (eBook)
https://doi.org/10.1007/978-3-032-06959-7



This Springer imprint is published by the registered company Springer Nature Switzerland AG
The registered company address is: Gewerbestrasse 11, 6330 Cham, Switzerland

# TC 289-DCM Members

Ing. Abiola Adebanjo, E-mail: abiola.adebanjo@uniosun.edu.ng
Prof. Karim Ait-Mokhtar, E-mail: karim.ait-mokhtar@univ-lr.fr
Mark G. Alexander, E-mail: mark.alexander@uct.ac.za
Ing. Ahmad Alhamdan, E-mail: alhamdana@cardiff.ac.uk
Mr. Babar Ali, E-mail: bali4@sheffield.ac.uk
Mr. Sofiane Amroun, E-mail: sxa1922@miami.edu
Dr. Carman Andrade, E-mail: carmen.andrade@upm.es
Prof. Jairo Andrade, E-mail: jairo.andrade@pucrs.br
Dr. Tushar Bansal, E-mail: tushar007success@gmail.com
Dr. Ana Baricevic, E-mail: abaricevic@grad.hr
Dr. Véronique Baroghel-Bouny, E-mail: veronique.baroghel-bouny@univ-eiffel.fr
Dr. María José Castillo, E-mail: mariajose.castillo@nexo.unnoba.edu.ar
Dr. Pedro Castro Borges, E-mail: pcb1963@gmail.com
Dr. Rachid Cherif, E-mail: rachid.cherif@univ-lr.fr
Dr. Gilberto Cidreira Keserle, E-mail: gilberto.cidreirakeserle@englobecorp.com
Prof. Patrick Dangla, E-mail: patrick.dangla@univ-eiffel.fr
Dr. Vinh Dao, E-mail: v.dao@uq.edu.au
Dr. Sufen Dong, E-mail: dongsufen@dlut.edu.cn
Prof. Yaghoob Farnam, E-mail: yf338@drexel.edu
Prof. Mette Rica Geiker, E-mail: mette.geiker@ntnu.no
Dr. Guoqing Geng, E-mail: ceegg@nus.edu.sg
Dr. Yan Geng, E-mail: gengyan11@163.com
Dr. Fabien Georget, E-mail: fabien.georget@epfl.ch
Ms. Ayana Ghosh, E-mail: a_ghosh@ce.iitr.ac.in
Dr. Tingyu Hao, E-mail: haotingyu@cribc.com
Ing. Ilhame Harbouz, E-mail: ilhame.harbouz@gmail.com
Dr. Karla Hornbostel, E-mail: karla.hornbostel@vegvesen.no
Dr. Ashlee Hossack, E-mail: ashlee.hossack@unb.ca
Mr. Md Mamun Hossain, E-mail: mhossain.cuet@gmail.com
Dr. Liming Huang, E-mail: lmhuang00@163.com
Dr. Fragkoulis Kanavaris, E-mail: kanavarisf@gmail.com

Dr. Hee Jeong Kim, E-mail: heejeong@arizona.edu
Dr. Kala Kondepudi, E-mail: sai_kala.kondepudi@tu-dresden.de
Prof. Konstantin Kovler, E-mail: cvrkost@cv.technion.ac.il
Ing. Marimuthu Krishnan, E-mail: marimuthusakthi2013@gmail.com
Dr. Gyanendra Kumar, E-mail: 2017rce9041@mnit.ac.in
Mr. Bowei Li, E-mail: li-bw20@mails.tsinghua.edu.cn
Prof. Kefei Li, E-mail: likefei@tsinghua.edu.cn (TC Chair)
Prof. Quanwang Li, E-mail: li_quanwang@mail.tsinghua.edu.cn
Dr. Ye Li, E-mail: ye.li@soton.ac.uk
Prof. Zhenming Li, E-mail: zhenmingli@hit.edu.cn
Prof. Chao Liu, E-mail: chaoliu@xauat.edu.cn
Prof. Xuemei Liu, E-mail: xuemei.liu@unimelb.edu.au
Mr. Jianxin Lu, E-mail: jxinlu@polyu.edu.hk
Dr. Riccardo Maddalena, E-mail: MaddalenaR@cardiff.ac.uk
Prof. José Fernando Martirena Hernández, E-mail: martirena@uclv.edu.cu
Mr. Ambokile Mwakyububu, E-mail: n.a.
Dr. Huu May Nguyen, E-mail: maynh@utt.edu.vn
Prof. Lars-Olof Nilsson, E-mail: lars-olof.nilsson@byggtek.lth.se
Ing. Nso Stephane Nyenti, E-mail: nyenti10@gmail.com
Dr. Malene Thostrup Pedersen, E-mail: malene.t.pedersen@ntnu.no
Prof. Radhakrishna G. Pillai, E-mail: pillai@iitm.ac.in
Dr. Shawn Platt, E-mail: shawn.platt@nist.gov
Mr. Chidchanok Plessudjai, E-mail: Chidchanok.plee@gmail.com
Dr. Gomasa Ramesh, E-mail: rameshgomasa1@gmail.com
Dr. Javier Sanchez Montero, E-mail: javier.sanchez@csic.es
Prof. Luca Sorelli, E-mail: luca.sorelli@gci.ulaval.ca
Dr. Henrik Erndahl Sørensen, E-mail: hks@dti.dk
Dr. Saarthak Surana, E-mail: SRNSAA001@myuct.ac.za
Mr. Siu Ming Wallace Tam, E-mail: smtam@cedd.gov.hk
Prof. Luping Tang, E-mail: tang.luping@chalmers.se
Prof. Michael Thomas, E-mail: mdat@unb.ca
Dr. Roberto J. Torrent, E-mail: torrent.concrete@gmail.com
Prof. Tamon Ueda, E-mail: ueda@eng.hokudai.ac.jp
Dr. André Valente Monteiro, E-mail: avmonteiro@lnec.pt
Prof. Talakokula Visalakshi, E-mail: basavishali@gmail.com
Dr. Junjie Wang, E-mail: junjiewang@tsinghua.edu.cn
Mr. Shengnian Wang, E-mail: wshengnian@cccc4.com
Mr. Hiu-Dan Wong, E-mail: hdwong@cedd.gov.hk
Dr. Min Wu, E-mail: mnwu@eng.au.dk
Dr. Meijie Xie, E-mail: majorthu2016@foxmail.com
Dr. Qingxiang Xiong, E-mail: qingxiangxiong@gmail.com
Mr. Haicheng Yang, E-mail: yanghaicheng@ccccltd.com
Dr. Jing Yu, E-mail: ceyujing@hku.hk
Mr. Andi Zahedi Rezzaieh, E-mail: azahe049@uottawa.ca
Dr. Junjie Zeng, E-mail: zjunjie1@cccc4.com (TC Deputy Chair)

Dr. Dongfang Zhang, E-mail: zhangdf8@mail.sysu.edu.cn
Prof. Mingzhong Zhang, E-mail: mingzhong.zhang@ucl.ac.uk
Prof. Peng Zhang, E-mail: zhp0221@163.com
Dr. Haiwei Zhu, E-mail: zhuhaiwei@nuaa.edu.cn
Dr. Rouzbeh Ziapour, E-mail: Rziap098@uottawa.ca

# RILEM TC 289-DCM

**STAR Contributors**

Abdelkarim Aït-Mokhtar, La Rochelle University, France

Carmen Andrade, International Center for Numerical Methods in Engineering (CIMNE), Spain

Mark Alexander, University of Cape Town, South Africa

Tushar Bansal, Sharda University, India

Véronique Baroghel-Bouny, Université Gustave Eiffel, France

Pedro Castro Borges, Centro de Investigación y de Estudios Avanzados del IPN, México

Rachid Cherif, La Rochelle University, France

Gilberto Cidreira Keserle, Englobe Corp., Canada

Alejandro Fernandez, Universidad Central de las Villas, Cuba

Mette Rica Geiker, Norwegian University of Science and Technology, Norway (Chap. 2 leader)

Guoqing Geng, National University of Singapore, Singapore

Tingyu Hao, Central Research Institute of Building and Construction, China

Liming Huang, Chalmers University of Technology, Sweden (Chap. 4 co-leader)

Fragkoulis Kanavaris, Arup Group Ltd., UK

Konstantin Kovler, National Building Research Institute, Israel

Bowei Li, Tsinghua University, China

Kefei Li, Tsinghua University, China (Chaps. 1, 5, 7 leader)

Quanwang Li, Tsinghua University, China

Xuemei Liu, The University of Melbourne, Australia

Riccardo Maddalena, Cardiff University, UK (Chap. 6 leader)

José Fernando Martirena-Hernández, Universidad Central de las Villas, Cuba

Henrik Erndahl Sørensen, Danish Technological Institute, Denmark

Javier Sanchez Montero, Eduardo Torroja Institute for Construction Sciences (IETcc), Spain

Saarthak Surana, University of Cape Town, South Africa

Luping Tang, Chalmers University of Technology, Sweden (Chap. 4 leader)

Roberto Torrent, Materials Advanced Services Ltd., Argentina

André Valente Monteiro, National Laboratory for Civil Engineering, Portugal
Talakokula Visalakshi, Mahindra University, India
Shengnian Wang, CCCC Fourth Harbor Engineering Co., Ltd., China
Haicheng Yang, CCCC Fourth Harbor Engineering Co., China
Junjie Zeng, CCCC Fourth Harbor Engineering Co., China (Chap. 3 leader)
Peng Zhang, Qingdao University of Technology, China

# RILEM Publications

The following list is presents the global offer of RILEM Publications, sorted by series. Each publication is available in printed version and/or in online version.

## RILEM Proceedings (PRO)

**PRO 1:** Durability of High Performance Concrete (ISBN: 2-912143-03-9; e-ISBN: 2-351580-12-5; e-ISBN: 2351580125); *Ed. H. Sommer*

**PRO 2:** Chloride Penetration into Concrete (ISBN: 2-912143-00-04; e-ISBN: 2912143454); *Eds. L.-O. Nilsson and J.-P. Ollivier*

**PRO 3:** Evaluation and Strengthening of Existing Masonry Structures (ISBN: 2-912143-02-0; e-ISBN: 2351580141); *Eds. L. Binda and C. Modena*

**PRO 4:** Concrete: From Material to Structure (ISBN: 2-912143-04-7; e-ISBN: 2351580206); *Eds. J.-P. Bournazel and Y. Malier*

**PRO 5:** The Role of Admixtures in High Performance Concrete (ISBN: 2-912143-05-5; e-ISBN: 2351580214); *Eds. J. G. Cabrera and R. Rivera-Villarreal*

**PRO 6:** High Performance Fiber Reinforced Cement Composites—HPFRCC 3 (ISBN: 2-912143-06-3; e-ISBN: 2351580222); *Eds. H. W. Reinhardt and A. E. Naaman*

**PRO 7:** 1st International RILEM Symposium on Self-Compacting Concrete (ISBN: 2-912143-09-8; e-ISBN: 2912143721); *Eds. Å. Skarendahl and Ö. Petersson*

**PRO 8:** International RILEM Symposium on Timber Engineering (ISBN: 2-912143-10-1; e-ISBN: 2351580230); *Ed. L. Boström*

**PRO 9:** 2nd International RILEM Symposium on Adhesion between Polymers and Concrete ISAP '99 (ISBN: 2-912143-11-X; e-ISBN: 2351580249); *Eds. Y. Ohama and M. Puterman*

**PRO 10:** 3rd International RILEM Symposium on Durability of Building and Construction Sealants (ISBN: 2-912143-13-6; e-ISBN: 2351580257); *Ed. A. T. Wolf*

**PRO 11:** 4th International RILEM Conference on Reflective Cracking in Pavements (ISBN: 2-912143-14-4; e-ISBN: 2351580265); *Eds. A. O. Abd El Halim, D. A. Taylor and El H. H. Mohamed*

**PRO 12:** International RILEM Workshop on Historic Mortars: Characteristics and Tests (ISBN: 2-912143-15-2; e-ISBN: 2351580273); *Eds. P. Bartos, C. Groot and J. J. Hughes*

**PRO 13:** 2nd International RILEM Symposium on Hydration and Setting (ISBN: 2-912143-16-0; e-ISBN: 2351580281); *Ed. A. Nonat*

**PRO 14:** Integrated Life-Cycle Design of Materials and Structures—ILCDES 2000 (ISBN: 951-758-408-3; e-ISBN: 235158029X); (ISSN: 0356-9403); *Ed. S. Sarja*

**PRO 15:** Fifth RILEM Symposium on Fibre-Reinforced Concretes (FRC)—BEFIB'2000 (ISBN: 2-912143-18-7; e-ISBN: 291214373X); *Eds. P. Rossi and G. Chanvillard*

**PRO 16:** Life Prediction and Management of Concrete Structures (ISBN: 2-912143-19-5; e-ISBN: 2351580303); *Ed. D. Naus*

**PRO 17:** Shrinkage of Concrete—Shrinkage 2000 (ISBN: 2-912143-20-9; e-ISBN: 2351580311); *Eds. V. Baroghel-Bouny and P.-C. Aïtcin*

**PRO 18:** Measurement and Interpretation of the On-Site Corrosion Rate (ISBN: 2-912143-21-7; e-ISBN: 235158032X); *Eds. C. Andrade, C. Alonso, J. Fullea, J. Polimon and J. Rodriguez*

**PRO 19:** Testing and Modelling the Chloride Ingress into Concrete (ISBN: 2-912143-22-5; e-ISBN: 2351580338); *Eds. C. Andrade and J. Kropp*

**PRO 20:** 1st International RILEM Workshop on Microbial Impacts on Building Materials (CD 02) (e-ISBN 978-2-35158-013-4); *Ed. M. Ribas Silva*

**PRO 21:** International RILEM Symposium on Connections between Steel and Concrete (ISBN: 2-912143-25-X; e-ISBN: 2351580346); *Ed. R. Eligehausen*

**PRO 22:** International RILEM Symposium on Joints in Timber Structures (ISBN: 2-912143-28-4; e-ISBN: 2351580354); *Eds. S. Aicher and H.-W. Reinhardt*

**PRO 23:** International RILEM Conference on Early Age Cracking in Cementitious Systems (ISBN: 2-912143-29-2; e-ISBN: 2351580362); *Eds. K. Kovler and A. Bentur*

**PRO 24:** 2nd International RILEM Workshop on Frost Resistance of Concrete (ISBN: 2-912143-30-6; e-ISBN: 2351580370); *Eds. M. J. Setzer, R. Auberg and H.-J. Keck*

**PRO 25:** International RILEM Workshop on Frost Damage in Concrete (ISBN: 2-912143-31-4; e-ISBN: 2351580389); *Eds. D. J. Janssen, M. J. Setzer and M. B. Snyder*

**PRO 26:** International RILEM Workshop on On-Site Control and Evaluation of Masonry Structures (ISBN: 2-912143-34-9; e-ISBN: 2351580141); *Eds. L. Binda and R. C. de Vekey*

**PRO 27:** International RILEM Symposium on Building Joint Sealants (CD03; e-ISBN: 235158015X); *Ed. A. T. Wolf*

**PRO 28:** 6th International RILEM Symposium on Performance Testing and Evaluation of Bituminous Materials—PTEBM'03 (ISBN: 2-912143-35-7; e-ISBN: 978-2-912143-77-8); *Ed. M. N. Partl*

**PRO 29:** 2nd International RILEM Workshop on Life Prediction and Ageing Management of Concrete Structures (ISBN: 2-912143-36-5; e-ISBN: 2912143780); *Ed. D. J. Naus*

**PRO 30:** 4th International RILEM Workshop on High Performance Fiber Reinforced Cement Composites—HPFRCC 4 (ISBN: 2-912143-37-3; e-ISBN: 2912143799); *Eds. A. E. Naaman and H. W. Reinhardt*

**PRO 31:** International RILEM Workshop on Test and Design Methods for Steel Fibre Reinforced Concrete: Background and Experiences (ISBN: 2-912143-38-1; e-ISBN: 2351580168); *Eds. B. Schnütgen and L. Vandewalle*

**PRO 32:** International Conference on Advances in Concrete and Structures 2 vol. (ISBN (set): 2-912143-41-1; e-ISBN: 2351580176); *Eds. Ying-shu Yuan, Surendra P. Shah and Heng-lin Lü*

**PRO 33:** 3rd International Symposium on Self-Compacting Concrete (ISBN: 2-912143-42-X; e-ISBN: 2912143713); *Eds. Ó. Wallevik and I. Níelsson*

**PRO 34:** International RILEM Conference on Microbial Impact on Building Materials (ISBN: 2-912143-43-8; e-ISBN: 2351580184); *Ed. M. Ribas Silva*

**PRO 35:** International RILEM TC 186-ISA on Internal Sulfate Attack and Delayed Ettringite Formation (ISBN: 2-912143-44-6; e-ISBN: 2912143802); *Eds. K. Scrivener and J. Skalny*

**PRO 36:** International RILEM Symposium on Concrete Science and Engineering—A Tribute to Arnon Bentur (ISBN: 2-912143-46-2; e-ISBN: 2912143586); *Eds. K. Kovler, J. Marchand, S. Mindess and J. Weiss*

**PRO 37:** 5th International RILEM Conference on Cracking in Pavements—Mitigation, Risk Assessment and Prevention (ISBN: 2-912143-47-0; e-ISBN: 2912143764); *Eds. C. Petit, I. Al-Qadi and A. Millien*

**PRO 38:** 3rd International RILEM Workshop on Testing and Modelling the Chloride Ingress into Concrete (ISBN: 2-912143-48-9; e-ISBN: 2912143578); *Eds. C. Andrade and J. Kropp*

**PRO 39:** 6th International RILEM Symposium on Fibre-Reinforced Concretes—BEFIB 2004 (ISBN: 2-912143-51-9; e-ISBN: 2912143748); *Eds. M. Di Prisco, R. Felicetti and G. A. Plizzari*

**PRO 40:** International RILEM Conference on the Use of Recycled Materials in Buildings and Structures (ISBN: 2-912143-52-7; e-ISBN: 2912143756); *Eds. E. Vázquez, Ch. F. Hendriks and G. M. T. Janssen*

**PRO 41:** RILEM International Symposium on Environment-Conscious Materials and Systems for Sustainable Development (ISBN: 2-912143-55-1; e-ISBN: 2912143640); *Eds. N. Kashino and Y. Ohama*

**PRO 42:** SCC'2005—China: 1st International Symposium on Design, Performance and Use of Self-Consolidating Concrete (ISBN: 2-912143-61-6; e-ISBN: 2912143624); *Eds. Zhiwu Yu, Caijun Shi, Kamal Henri Khayat and Youjun Xie*

**PRO 43:** International RILEM Workshop on Bonded Concrete Overlays (e-ISBN: 2-912143-83-7); *Eds. J. L. Granju and J. Silfwerbrand*

**PRO 44:** 2nd International RILEM Workshop on Microbial Impacts on Building Materials (CD11) (e-ISBN: 2-912143-84-5); *Ed. M. Ribas Silva*

**PRO 45:** 2nd International Symposium on Nanotechnology in Construction, Bilbao (ISBN: 2-912143-87-X; e-ISBN: 2912143888); *Eds. Peter J. M. Bartos, Yolanda de Miguel and Antonio Porro*

**PRO 46:** Concrete Life'06—International RILEM-JCI Seminar on Concrete Durability and Service Life Planning: Curing, Crack Control, Performance in Harsh Environments (ISBN: 2-912143-89-6; e-ISBN: 291214390X); *Ed. K. Kovler*

**PRO 47:** International RILEM Workshop on Performance Based Evaluation and Indicators for Concrete Durability (ISBN: 978-2-912143-95-2; e-ISBN: 9782912143969); *Eds. V. Baroghel-Bouny, C. Andrade, R. Torrent and K. Scrivener*

**PRO 48:** 1st International RILEM Symposium on Advances in Concrete through Science and Engineering (e-ISBN: 2-912143-92-6); *Eds. J. Weiss, K. Kovler, J. Marchand, and S. Mindess*

**PRO 49:** International RILEM Workshop on High Performance Fiber Reinforced Cementitious Composites in Structural Applications (ISBN: 2-912143-93-4; e-ISBN: 2912143942); *Eds. G. Fischer and V. C. Li*

**PRO 50:** 1st International RILEM Symposium on Textile Reinforced Concrete (ISBN: 2-912143-97-7; e-ISBN: 2351580087); *Eds. Josef Hegger, Wolfgang Brameshuber and Norbert Will*

**PRO 51:** 2nd International Symposium on Advances in Concrete through Science and Engineering (ISBN: 2-35158-003-6; e-ISBN: 2-35158-002-8); *Eds. J. Marchand, B. Bissonnette, R. Gagné, M. Jolin and F. Paradis*

**PRO 52:** Volume Changes of Hardening Concrete: Testing and Mitigation (ISBN: 2-35158-004-4; e-ISBN: 2-35158-005-2); *Eds. O. M. Jensen, P. Lura and K. Kovler*

**PRO 53:** High Performance Fiber Reinforced Cement Composites—HPFRCC5 (ISBN: 978-2-35158-046-2; e-ISBN: 978-2-35158-089-9); *Eds. H. W. Reinhardt and A. E. Naaman*

**PRO 54:** 5th International RILEM Symposium on Self-Compacting Concrete (ISBN: 978-2-35158-047-9; e-ISBN: 978-2-35158-088-2); *Eds. G. De Schutter and V. Boel*

**PRO 55:** International RILEM Symposium Photocatalysis, Environment and Construction Materials (ISBN: 978-2-35158-056-1; e-ISBN: 978-2-35158-057-8); *Eds. P. Baglioni and L. Cassar*

**PRO 56:** International RILEM Workshop on Integral Service Life Modelling of Concrete Structures (ISBN 978-2-35158-058-5; e-ISBN: 978-2-35158-090-5); *Eds. R. M. Ferreira, J. Gulikers and C. Andrade*

**PRO 57:** RILEM Workshop on Performance of cement-based materials in aggressive aqueous environments (e-ISBN: 978-2-35158-059-2); *Ed. N. De Belie*

**PRO 58:** International RILEM Symposium on Concrete Modelling—CONMOD'08 (ISBN: 978-2-35158-060-8; e-ISBN: 978-2-35158-076-9); *Eds. E. Schlangen and G. De Schutter*

**PRO 59:** International RILEM Conference on On Site Assessment of Concrete, Masonry and Timber Structures—SACoMaTiS 2008 (ISBN set: 978-2-35158-061-5; e-ISBN: 978-2-35158-075-2); *Eds. L. Binda, M. di Prisco and R. Felicetti*

**PRO 60:** Seventh RILEM International Symposium on Fibre Reinforced Concrete: Design and Applications—BEFIB 2008 (ISBN: 978-2-35158-064-6; e-ISBN: 978-2-35158-086-8); *Ed. R. Gettu*

**PRO 61:** 1st International Conference on Microstructure Related Durability of Cementitious Composites 2 vol., (ISBN: 978-2-35158-065-3; e-ISBN: 978-2-35158-084-4); *Eds. W. Sun, K. van Breugel, C. Miao, G. Ye and H. Chen*

**PRO 62:** NSF/ RILEM Workshop: In-situ Evaluation of Historic Wood and Masonry Structures (e-ISBN: 978-2-35158-068-4); *Eds. B. Kasal, R. Anthony and M. Drdácký*

**PRO 63:** Concrete in Aggressive Aqueous Environments: Performance, Testing and Modelling, 2 vol., (ISBN: 978-2-35158-071-4; e-ISBN: 978-2-35158-082-0); *Eds. M. G. Alexander and A. Bertron*

**PRO 64:** Long Term Performance of Cementitious Barriers and Reinforced Concrete in Nuclear Power Plants and Waste Management—NUCPERF 2009 (ISBN: 978-2-35158-072-1; e-ISBN: 978-2-35158-087-5); *Eds. V. L'Hostis, R. Gens and C. Gallé*

**PRO 65:** Design Performance and Use of Self-consolidating Concrete—SCC'2009 (ISBN: 978-2-35158-073-8; e-ISBN: 978-2-35158-093-6); *Eds. C. Shi, Z. Yu, K. H. Khayat and P. Yan*

**PRO 66:** 2nd International RILEM Workshop on Concrete Durability and Service Life Planning—ConcreteLife'09 (ISBN: 978-2-35158-074-5; ISBN: 978-2-35158-074-5); *Ed. K. Kovler*

**PRO 67**: Repairs Mortars for Historic Masonry (e-ISBN: 978-2-35158-083-7); *Ed. C. Groot*

**PRO 68:** Proceedings of the 3rd International RILEM Symposium on 'Rheology of Cement Suspensions such as Fresh Concrete (ISBN 978-2-35158-091-2; e-ISBN: 978-2-35158-092-9); *Eds. O. H. Wallevik, S. Kubens and S. Oesterheld*

**PRO 69:** 3rd International PhD Student Workshop on 'Modelling the Durability of Reinforced Concrete (ISBN: 978-2-35158-095-0); *Eds. R. M. Ferreira, J. Gulikers and C. Andrade*

**PRO 70:** 2nd International Conference on 'Service Life Design for Infrastructure' (ISBN set: 978-2-35158-096-7, e-ISBN: 978-2-35158-097-4); *Eds. K. van Breugel, G. Ye and Y. Yuan*

**PRO 71:** Advances in Civil Engineering Materials—The 50-year Teaching Anniversary of Prof. Sun Wei' (ISBN: 978-2-35158-098-1; e-ISBN: 978-2-35158-099-8); *Eds. C. Miao, G. Ye and H. Chen*

**PRO 72:** First International Conference on 'Advances in Chemically-Activated Materials—CAM'2010' (2010), 264 pp., ISBN: 978-2-35158-101-8; e-ISBN: 978-2-35158-115-5; *Eds. Caijun Shi and Xiaodong Shen*

**PRO 73:** 2nd International Conference on 'Waste Engineering and Management—ICWEM 2010' (2010), 894 pp., ISBN: 978-2-35158-102-5; e-ISBN: 978-2-35158-103-2, *Eds. J. Zh. Xiao, Y. Zhang, M. S. Cheung and R. Chu*

**PRO 74:** International RILEM Conference on 'Use of Superabsorbent Polymers and Other New Additives in Concrete' (2010) 374 pp., ISBN: 978-2-35158-104-9; e-ISBN: 978-2-35158-105-6; *Eds. O.M. Jensen, M.T. Hasholt, and S. Laustsen*

**PRO 75:** International Conference on 'Material Science—2nd ICTRC—Textile Reinforced Concrete—Theme 1' (2010) 436 pp., ISBN: 978-2-35158-106-3; e-ISBN: 978-2-35158-107-0; *Ed. W. Brameshuber*

**PRO 76:** International Conference on 'Material Science—HetMat—Modelling of Heterogeneous Materials—Theme 2' (2010) 255 pp., ISBN: 978-2-35158-108-7; e-ISBN: 978-2-35158-109-4; *Ed. W. Brameshuber*

**PRO 77:** International Conference on 'Material Science—AdIPoC—Additions Improving Properties of Concrete—Theme 3' (2010) 459 pp., ISBN: 978-2-35158-110-0; e-ISBN: 978-2-35158-111-7; *Ed. W. Brameshuber*

**PRO 78:** 2nd Historic Mortars Conference and RILEM TC 203-RHM Final Workshop—HMC2010 (2010) 1416 pp., e-ISBN: 978-2-35158-112-4; *Eds. J. Válek, C. Groot and J. J. Hughes*

**PRO 79:** International RILEM Conference on Advances in Construction Materials Through Science and Engineering (2011) 213 pp., ISBN: 978-2-35158-116-2, e-ISBN: 978-2-35158-117-9; *Eds. Christopher Leung and K.T. Wan*

**PRO 80:** 2nd International RILEM Conference on Concrete Spalling due to Fire Exposure (2011) 453 pp., ISBN: 978-2-35158-118-6; e-ISBN: 978-2-35158-119-3; *Eds. E.A.B. Koenders and F. Dehn*

**PRO 81:** 2nd International RILEM Conference on Strain Hardening Cementitious Composites (SHCC2-Rio) (2011) 451 pp., ISBN: 978-2-35158-120-9; e-ISBN: 978-2-35158-121-6; *Eds. R.D. Toledo Filho, F.A. Silva, E.A.B. Koenders and E.M.R. Fairbairn*

**PRO 82:** 2nd International RILEM Conference on Progress of Recycling in the Built Environment (2011) 507 pp., e-ISBN: 978-2-35158-122-3; *Eds. V.M. John, E. Vazquez, S.C. Angulo and C. Ulsen*

**PRO 83:** 2nd International Conference on Microstructural-related Durability of Cementitious Composites (2012) 250 pp., ISBN: 978-2-35158-129-2; e-ISBN: 978-2-35158-123-0; *Eds. G. Ye, K. van Breugel, W. Sun and C. Miao*

**PRO 84:** CONSEC13—Seventh International Conference on Concrete under Severe Conditions—Environment and Loading (2013) 1930 pp., ISBN: 978-2-35158-124-7; e-ISBN: 978-2- 35158-134-6; *Eds. Z.J. Li, W. Sun, C.W. Miao, K. Sakai, O.E. Gjorv and N. Banthia*

**PRO 85:** RILEM-JCI International Workshop on Crack Control of Mass Concrete and Related issues concerning Early-Age of Concrete Structures—ConCrack 3—Control of Cracking in Concrete Structures 3 (2012) 237 pp., ISBN: 978-2-35158-125-4; e-ISBN: 978-2-35158-126-1; *Eds. F. Toutlemonde and J.-M. Torrenti*

**PRO 86:** International Symposium on Life Cycle Assessment and Construction (2012) 414 pp., ISBN: 978-2-35158-127-8, e-ISBN: 978-2-35158-128-5; *Eds. A. Ventura and C. de la Roche*

**PRO 87:** UHPFRC 2013—RILEM-fib-AFGC International Symposium on Ultra-High Performance Fibre-Reinforced Concrete (2013), ISBN: 978-2-35158-130-8, e-ISBN: 978-2-35158-131-5; *Eds. F. Toutlemonde*

**PRO 88:** 8th RILEM International Symposium on Fibre Reinforced Concrete (2012) 344 pp., ISBN: 978-2-35158-132-2; e-ISBN: 978-2-35158-133-9; *Eds. Joaquim A.O. Barros*

**PRO 89:** RILEM International workshop on performance-based specification and control of concrete durability (2014) 678 pp., ISBN: 978-2-35158-135-3; e-ISBN: 978-2-35158-136-0; *Eds. D. Bjegović, H. Beushausen and M. Serdar*

**PRO 90:** 7th RILEM International Conference on Self-Compacting Concrete and of the 1st RILEM International Conference on Rheology and Processing of Construction Materials (2013) 396 pp., ISBN: 978-2-35158-137-7; e-ISBN: 978-2-35158-138-4; *Eds. Nicolas Roussel and Hela Bessaies-Bey*

**PRO 91:** CONMOD 2014—RILEM International Symposium on Concrete Modelling (2014), ISBN: 978-2-35158-139-1; e-ISBN: 978-2-35158-140-7; *Eds. Kefei Li, Peiyu Yan and Rongwei Yang*

**PRO 92:** CAM 2014—2nd International Conference on advances in chemically-activated materials (2014) 392 pp., ISBN: 978-2-35158-141-4; e-ISBN: 978-2-35158-142-1; *Eds. Caijun Shi and Xiadong Shen*

**PRO 93:** SCC 2014—3rd International Symposium on Design, Performance and Use of Self-Consolidating Concrete (2014) 438 pp., ISBN: 978-2-35158-143-8; e-ISBN: 978-2-35158-144-5; *Eds. Caijun Shi, Zhihua Ou and Kamal H. Khayat*

**PRO 94 (online version):** HPFRCC-7—7th RILEM conference on High performance fiber reinforced cement composites (2015), e-ISBN: 978-2-35158-146-9; *Eds. H.W. Reinhardt, G.J. Parra-Montesinos and H. Garrecht*

**PRO 95:** International RILEM Conference on Application of superabsorbent polymers and other new admixtures in concrete construction (2014), ISBN: 978-2-35158-147-6; e-ISBN: 978-2-35158-148-3; *Eds. Viktor Mechtcherine and Christof Schroefl*

**PRO 96 (online version):** XIII DBMC: XIII International Conference on Durability of Building Materials and Components (2015), e-ISBN: 978-2-35158-149-0; *Eds. M. Quattrone and V.M. John*

**PRO 97:** SHCC3—3rd International RILEM Conference on Strain Hardening Cementitious Composites (2014), ISBN: 978-2-35158-150-6; e-ISBN: 978-2-35158-151-3; *Eds. E. Schlangen, M.G. Sierra Beltran, M. Lukovic and G. Ye*

**PRO 98:** FERRO-11—11th International Symposium on Ferrocement and 3rd ICTRC—International Conference on Textile Reinforced Concrete (2015), ISBN: 978-2-35158-152-0; e-ISBN: 978-2-35158-153-7; *Ed. W. Brameshuber*

**PRO 99 (online version):** ICBBM 2015—1st International Conference on Bio-Based Building Materials (2015), e-ISBN: 978-2-35158-154-4; *Eds. S. Amziane and M. Sonebi*

**PRO 100:** SCC16—RILEM Self-Consolidating Concrete Conference (2016), ISBN: 978-2-35158-156-8; e-ISBN: 978-2-35158-157-5; *Ed. Kamal H. Kayat*

**PRO 101 (online version):** III Progress of Recycling in the Built Environment (2015), e-ISBN: 978-2-35158-158-2; *Eds I. Martins, C. Ulsen and S. C. Angulo*

**PRO 102 (online version):** RILEM Conference on Microorganisms-Cementitious Materials Interactions (2016), e-ISBN: 978-2-35158-160-5; *Eds. Alexandra Bertron, Henk Jonkers and Virginie Wiktor*

**PRO 103 (online version):** ACESC'16—Advances in Civil Engineering and Sustainable Construction (2016), e-ISBN: 978-2-35158-161-2; *Eds. T.Ch. Madhavi, G. Prabhakar, Santhosh Ram and P.M. Rameshwaran*

**PRO 104 (online version):** SSCS'2015—Numerical Modeling—Strategies for Sustainable Concrete Structures (2015), e-ISBN: 978-2-35158-162-9

**PRO 105:** 1st International Conference on UHPC Materials and Structures (2016), ISBN: 978-2-35158-164-3; e-ISBN: 978-2-35158-165-0

**PRO 106:** AFGC-ACI-fib-RILEM International Conference on Ultra-High-Performance Fibre-Reinforced Concrete—UHPFRC 2017 (2017), ISBN: 978-2-35158-166-7; e-ISBN: 978-2-35158-167-4; *Eds. François Toutlemonde and Jacques Resplendino*

**PRO 107 (online version):** XIV DBMC—14th International Conference on Durability of Building Materials and Components (2017), e-ISBN: 978-2-35158-159-9; *Eds. Geert De Schutter, Nele De Belie, Arnold Janssens and Nathan Van Den Bossche*

**PRO 108:** MSSCE 2016—Innovation of Teaching in Materials and Structures (2016), ISBN: 978-2-35158-178-0; e-ISBN: 978-2-35158-179-7; *Ed. Per Goltermann*

**PRO 109 (2 volumes):** MSSCE 2016—Service Life of Cement-Based Materials and Structures (2016), ISBN Vol. 1: 978-2-35158-170-4; Vol. 2: 978-2-35158-171-4; Set Vol. 1&2: 978-2-35158-172-8; e-ISBN: 978-2-35158-173-5; *Eds. Miguel Azenha, Ivan Gabrijel, Dirk Schlicke, Terje Kanstad and Ole Mejlhede Jensen*

**PRO 110:** MSSCE 2016—Historical Masonry (2016), ISBN: 978-2-35158-178-0; e-ISBN: 978-2-35158-179-7; *Eds. Inge Rörig-Dalgaard and Ioannis Ioannou*

**PRO 111:** MSSCE 2016—Electrochemistry in Civil Engineering (2016); ISBN: 978-2-35158-176-6; e-ISBN: 978-2-35158-177-3; *Ed. Lisbeth M. Ottosen*

**PRO 112:** MSSCE 2016—Moisture in Materials and Structures (2016), ISBN: 978-2-35158-178-0; e-ISBN: 978-2-35158-179-7; *Eds. Kurt Kielsgaard Hansen, Carsten Rode and Lars-Olof Nilsson*

**PRO 113:** MSSCE 2016—Concrete with Supplementary Cementitious Materials (2016), ISBN: 978-2-35158-178-0; e-ISBN: 978-2-35158-179-7; *Eds. Ole Mejlhede Jensen, Konstantin Kovler and Nele De Belie*

**PRO 114:** MSSCE 2016—Frost Action in Concrete (2016), ISBN: 978-2-35158-182-7; e-ISBN: 978-2-35158-183-4; *Eds. Marianne Tange Hasholt, Katja Fridh and R. Doug Hooton*

**PRO 115:** MSSCE 2016—Fresh Concrete (2016), ISBN: 978-2-35158-184-1; e-ISBN: 978-2-35158-185-8; *Eds. Lars N. Thrane, Claus Pade, Oldrich Svec and Nicolas Roussel*

**PRO 116:** BEFIB 2016—9th RILEM International Symposium on Fiber Reinforced Concrete (2016), ISBN: 978-2-35158-187-2; e-ISBN: 978-2-35158-186-5; *Eds. N. Banthia, M. di Prisco and S. Soleimani-Dashtaki*

**PRO 117:** 3rd International RILEM Conference on Microstructure Related Durability of Cementitious Composites (2016), ISBN: 978-2-35158-188-9; e-ISBN: 978-2-35158-189-6; *Eds. Changwen Miao, Wei Sun, Jiaping Liu, Huisu Chen, Guang Ye and Klaas van Breugel*

**PRO 118 (4 volumes):** International Conference on Advances in Construction Materials and Systems (2017), ISBN Set: 978-2-35158-190-2; Vol. 1: 978-2-35158-193-3; Vol. 2: 978-2-35158-194-0; Vol. 3: ISBN:978-2-35158-195-7; Vol. 4: ISBN:978-2-35158-196-4; e-ISBN: 978-2-35158-191-9; *Ed. Manu Santhanam*

**PRO 119 (online version):** ICBBM 2017—Second International RILEM Conference on Bio-based Building Materials, (2017), e-ISBN: 978-2-35158-192-6; *Ed. Sofiane Amziane*

**PRO 120 (2 volumes):** EAC-02—2nd International RILEM/COST Conference on Early Age Cracking and Serviceability in Cement-based Materials and Structures, (2017), Vol. 1: 978-2-35158-199-5, Vol. 2: 978-2-35158-200-8, Set: 978-2-35158-197-1, e-ISBN: 978-2-35158-198-8; *Eds. Stéphanie Staquet and Dimitrios Aggelis*

**PRO 121 (2 volumes):** SynerCrete18: Interdisciplinary Approaches for Cement-based Materials and Structural Concrete: Synergizing Expertise and Bridging Scales of Space and Time, (2018), Set: 978-2-35158-202-2, Vol.1: 978-2-35158-211-4, Vol.2: 978-2-35158-212-1, e-ISBN: 978-2-35158-203-9; *Eds. Miguel Azenha, Dirk Schlicke, Farid Benboudjema, Agnieszka Knoppik*

**PRO 122:** SCC'2018 China—Fourth International Symposium on Design, Performance and Use of Self-Consolidating Concrete, (2018), ISBN: 978-2-35158-204-6, e-ISBN: 978-2-35158-205-3; *Eds. C. Shi, Z. Zhang, K. H. Khayat*

**PRO 123:** Final Conference of RILEM TC 253-MCI: Microorganisms-Cementitious Materials Interactions (2018), Set: 978-2-35158-207-7, Vol.1: 978-2-35158-209-1, Vol.2: 978-2-35158-210-7, e-ISBN: 978-2-35158-206-0; *Ed. Alexandra Bertron*

**PRO 124 (online version):** Fourth International Conference Progress of Recycling in the Built Environment (2018), e-ISBN: 978-2-35158-208-4; *Eds. Isabel M. Martins, Carina Ulsen, Yury Villagran*

**PRO 125 (online version):** SLD4—4th International Conference on Service Life Design for Infrastructures (2018), e-ISBN: 978-2-35158-213-8; *Eds. Guang Ye, Yong Yuan, Claudia Romero Rodriguez, Hongzhi Zhang, Branko Savija*

**PRO 126:** Workshop on Concrete Modelling and Material Behaviour in honor of Professor Klaas van Breugel (2018), ISBN: 978-2-35158-214-5, e-ISBN: 978-2-35158-215-2; *Ed. Guang Ye*

**PRO 127 (online version):** CONMOD2018—Symposium on Concrete Modelling (2018), e-ISBN: 978-2-35158-216-9; *Eds. Erik Schlangen, Geert de Schutter, Branko Savija, Hongzhi Zhang, Claudia Romero Rodriguez*

**PRO 128:** SMSS2019—International Conference on Sustainable Materials, Systems and Structures (2019), ISBN: 978-2-35158-217-6, e-ISBN: 978-2-35158-218-3

**PRO 129:** 2nd International Conference on UHPC Materials and Structures (UHPC2018-China), ISBN: 978-2-35158-219-0, e-ISBN: 978-2-35158-220-6

**PRO 130:** 5th Historic Mortars Conference (2019), ISBN: 978-2-35158-221-3, e-ISBN: 978-2-35158-222-0; *Eds. José Ignacio Álvarez, José María Fernández, Íñigo Navarro, Adrián Durán, Rafael Sirera*

**PRO 131 (online version):** 3rd International Conference on Bio-Based Building Materials (ICBBM2019), e-ISBN: 978-2-35158-229-9; *Eds. Mohammed Sonebi, Sofiane Amziane, Jonathan Page*

**PRO 132:** IRWRMC'18—International RILEM Workshop on Rheological Measurements of Cement-based Materials (2018), ISBN: 978-2-35158-230-5, e-ISBN: 978-2-35158-231-2; *Eds. Chafika Djelal, Yannick Vanhove*

**PRO 133 (online version):** CO2STO2019—International Workshop CO2 Storage in Concrete (2019), e-ISBN: 978-2-35158-232-9; *Eds. Assia Djerbi, Othman Omikrine-Metalssi, Teddy Fen-Chong*

**PRO 134:** 3rd ACF/HNU International Conference on UHPC Materials and Structures—UHPC'2020, ISBN: 978-2-35158-233-6, e-ISBN: 978-2-35158-234-3; *Eds. Caijun Shi and Jiaping Liu*

**PRO 135:** Fourth International Conference on Chemically Activated Materials (CAM2021), ISBN: 978-2-35158-235-0, e-ISBN: 978-2-35158-236-7; *Eds. Caijun Shi and Xiang Hu*

**PRO 136:** Lime Based Materials for Repairing Historic Structures (2022), ISBN: 978-2-35158-240-4, e-ISBN: 978-2-35158-237-4; *Eds. Ioanna Papayianni and Jan Válek*

**PRO 137 (online version):** V International Conference Progress of Recycling in the Built Environment (2023), e-ISBN: 978-2-35158-238-1; *Eds. Ines Döring, Yury Villagrán Zaccardi, Anette Müller*

## RILEM Reports (REP)

**Report 19:** Considerations for Use in Managing the Aging of Nuclear Power Plant Concrete Structures (ISBN: 2-912143-07-1); *Ed. D. J. Naus*

**Report 20:** Engineering and Transport Properties of the Interfacial Transition Zone in Cementitious Composites (ISBN: 2-912143-08-X); *Eds. M. G. Alexander, G. Arliguie, G. Ballivy, A. Bentur and J. Marchand*

**Report 21:** Durability of Building Sealants (ISBN: 2-912143-12-8); *Ed. A. T. Wolf*

**Report 22:** Sustainable Raw Materials—Construction and Demolition Waste (ISBN: 2-912143-17-9); *Eds. C. F. Hendriks and H. S. Pietersen*

**Report 23:** Self-Compacting Concrete state-of-the-art report (ISBN: 2-912143-23-3); *Eds. Å. Skarendahl and Ö. Petersson*

**Report 24:** Workability and Rheology of Fresh Concrete: Compendium of Tests (ISBN: 2-912143-32-2); *Eds. P. J. M. Bartos, M. Sonebi and A. K. Tamimi*

**Report 25:** Early Age Cracking in Cementitious Systems (ISBN: 2-912143-33-0); *Ed. A. Bentur*

**Report 26:** Towards Sustainable Roofing (Joint Committee CIB/RILEM) (CD 07) (e-ISBN 978-2-912143-65-5); *Eds. Thomas W. Hutchinson and Keith Roberts*

**Report 27:** Condition Assessment of Roofs (Joint Committee CIB/RILEM) (CD 08) (e-ISBN 978-2-912143-66-2); *Ed. CIB W 83/RILEM TC166-RMS*

**Report 28:** Final report of RILEM TC 167-COM 'Characterisation of Old Mortars with Respect to Their Repair (ISBN: 978-2-912143-56-3); *Eds. C. Groot, G. Ashall and J. Hughes*

**Report 29:** Pavement Performance Prediction and Evaluation (PPPE): Interlaboratory Tests (e-ISBN: 2-912143-68-3); *Eds. M. Partl and H. Piber*

**Report 30:** Final Report of RILEM TC 198-URM 'Use of Recycled Materials' (ISBN: 2-912143-82-9; e-ISBN: 2-912143-69-1); *Eds. Ch. F. Hendriks, G. M. T. Janssen and E. Vázquez*

**Report 31:** Final Report of RILEM TC 185-ATC 'Advanced testing of cement-based materials during setting and hardening' (ISBN: 2-912143-81-0; e-ISBN: 2-912143-70-5); *Eds. H. W. Reinhardt and C. U. Grosse*

**Report 32:** Probabilistic Assessment of Existing Structures. A JCSS publication (ISBN 2-912143-24-1); *Ed. D. Diamantidis*

**Report 33:** State-of-the-Art Report of RILEM Technical Committee TC 184-IFE 'Industrial Floors' (ISBN 2-35158-006-0); *Ed. P. Seidler*

**Report 34:** Report of RILEM Technical Committee TC 147-FMB 'Fracture mechanics applications to anchorage and bond' Tension of Reinforced Concrete Prisms—Round Robin Analysis and Tests on Bond (e-ISBN 2-912143-91-8); *Eds. L. Elfgren and K. Noghabai*

**Report 35:** Final Report of RILEM Technical Committee TC 188-CSC 'Casting of Self Compacting Concrete' (ISBN 2-35158-001-X; e-ISBN: 2-912143-98-5); *Eds. Å. Skarendahl and P. Billberg*

**Report 36:** State-of-the-Art Report of RILEM Technical Committee TC 201-TRC 'Textile Reinforced Concrete' (ISBN 2-912143-99-3); *Ed. W. Brameshuber*

**Report 37:** State-of-the-Art Report of RILEM Technical Committee TC 192-ECM 'Environment-conscious construction materials and systems' (ISBN: 978-2-35158-053-0); *Eds. N. Kashino, D. Van Gemert and K. Imamoto*

**Report 38:** State-of-the-Art Report of RILEM Technical Committee TC 205-DSC 'Durability of Self-Compacting Concrete' (ISBN: 978-2-35158-048-6); *Eds. G. De Schutter and K. Audenaert*

**Report 39:** Final Report of RILEM Technical Committee TC 187-SOC 'Experimental determination of the stress-crack opening curve for concrete in tension' (ISBN 978-2-35158-049-3); *Ed. J. Planas*

**Report 40:** State-of-the-Art Report of RILEM Technical Committee TC 189-NEC 'Non-Destructive Evaluation of the Penetrability and Thickness of the Concrete Cover' (ISBN 978-2-35158-054-7); *Eds. R. Torrent and L. Fernández Luco*

**Report 41:** State-of-the-Art Report of RILEM Technical Committee TC 196-ICC 'Internal Curing of Concrete' (ISBN 978-2-35158-009-7); *Eds. K. Kovler and O. M. Jensen*

**Report 42:** 'Acoustic Emission and Related Non-destructive Evaluation Techniques for Crack Detection and Damage Evaluation in Concrete'—Final Report of RILEM Technical Committee 212-ACD (e-ISBN: 978-2-35158-100-1); *Ed. M. Ohtsu*

**Report 45:** Repair Mortars for Historic Masonry—State-of-the-Art Report of RILEM Technical Committee TC 203-RHM (e-ISBN: 978-2-35158-163-6); *Eds. Paul Maurenbrecher and Caspar Groot*

**Report 46:** Surface delamination of concrete industrial floors and other durability related aspects guide—Report of RILEM Technical Committee TC 268-SIF (e-ISBN: 978-2-35158-201-5); *Ed. Valerie Pollet*

# Contents

# Abbreviations

| | |
|---|---|
| ATM | Atmospheric exposure zone |
| CAH | Calcium Aluminate Hydrates |
| CASH | Calcium Aluminate Silicate Hydrates |
| Cement | Binder with cementitious properties |
| CH | Calcium Hydroxide (cement hydrates) |
| C-S-H | Calcium Silicate Hydrates |
| EPMA | Electron Probe Micro Analyzer |
| FA | Fly Ash |
| GRBA | Ground Rice husk-Bark Ash |
| ITZ | Interfacial Transition Zone (between cement paste and aggregates) |
| LC3 | Calcined Clay Limestone Cements |
| LPR | Linear Polarization Resistance |
| LWA | Lightweight Aggregates |
| OPC | Ordinary Portland Cement (Concrete) |
| SCM | Supplementary (secondary) Cementitious Materials |
| SEM | Scanning Electronic Microscopy |
| SF | Silica Fume |
| SPL | Splashing exposure zone |
| SUB | Submerged exposure zone |
| TID | Tidal exposure zone |
| ToW | Time of Wetness |
| UPV | Ultrasonic Pulse Velocity |
| w/c (w/b) ratio | Water to cement (binder) ratio |
| XRD | X-Ray Diffraction |
| μXRF | Micro X-Ray Fluorescence |

# Chapter 1
# Introduction

**Kefei Li, Pedro Castro Borges, Mark Alexander, Saarthak Surana, Abdelkarim Aït-Mokhtar, Rachid Cherif, José Fernando Martirena-Hernández, André Valente Monteiro, Carmen Andrade, Luping Tang, Henrik Erndahl Sørensen, and Mette Rica Geiker**

K. Li (✉)
Civil Engineering Department, Tsinghua University, Beijing, China
e-mail: likefei@tsinghua.edu.cn

P. C. Borges
Centro de Investigación y de Estudios Avanzados del IPN, Unidad Mérida, Mérida, Yucatán, México

M. Alexander · S. Surana
Department of Civil Engineering, University of Cape Town, Cape Town, South Africa

A. Aït-Mokhtar · R. Cherif
LaSIE UMR CNRS 7356, La Rochelle University, La Rochelle Cedex 1, France

J. F. Martirena-Hernández
Facultad de Construcciones, Universidad Central de las Villas, Santa Clara, Villa Clara, Cuba

A. V. Monteiro
National Laboratory for Civil Engineering, Lisbon, Portugal

C. Andrade
International Center for Numerical Methods in Engineering (CIMNE), Barcelona, Spain

L. Tang
Department of Architecture and Civil Engineering, Chalmers University of Technology, Gothenburg, Sweden

H. E. Sørensen
Building and Construction, Danish Technological Institute, Taastrup, Denmark

M. R. Geiker
Department of Structural Engineering, Norwegian University of Science and Technology, Trondheim, Norway

 
K. Li et al. (eds.), *Long-Term Performance of Structural Concretes in Marine Exposure Conditions*, RILEM State-of-the-Art Reports Volume,
https://doi.org/10.1007/978-3-032-06959-7_1

## 1.1 Background

Nowadays, structural concretes are required to fulfil their expected performance during long service lives, e.g. 50 years for buildings and more than 100 years for large-scale infrastructures [14]. The knowledge on their long-term durability is crucial for the design and management of these structures and infrastructures, especially for those exposed in aggressive environments such as the marine environment. The durability performance of structural concretes in marine environments is highly complex, involving multiple actions such as chloride ingress, carbonation, freeze–thaw, seawater erosion and their coupled effects [3]. The marine exposure stations/sites provide unique data sources to investigate the long-term performance of concretes. These data also support the design, assessment and prediction of long-term durability of concrete structures and infrastructures. That is the very reason that the engineering community values the practice of long-term field exposure.

Since the 1930s, the long-term exposure practice has been engaged in engineering practice and the obtained knowledge can be witnessed by the extended design working lives for concrete structures exposed in severe marine environments. Despite these achievements, the exploitation of long-term field data is not yet satisfactory, and the added value of these data has not been fully generated. The data collection from exposure stations is rather intuitive, and a systematic format for data collection/presentation is missing; more deepened interpretation of these data, obtained through both labor and budget demanding procedures, are expected to provide more information for durability performance; and more formalized guideline is expected for the operation of exposure stations and the support for life-cycle management of concrete structures and infrastructures. Aiming at these lacks, the RILEM TC 289-DCM is committed to providing solutions and tools for these needs. This STAR gives a comprehensive review on state-of-the-art of the deterioration mechanisms of concretes in marine environments, the presentation of long-term filed exposure data, the interpretation and modelling from these data, and the exploitation of these data. From the return of experiences of major reviewed exposure sites, it is expected that the practice of field exposure in marine environments can be shared in wider scope for both research and practical purposes.

## 1.2 Major Exposure Sites

### *1.2.1 Exposure Sites in America*

The United States have commenced the systematic practice of long-term exposure of concrete materials in marine environments since 1930s, building exposure stations in four typical marine climates in the coastal areas [21, 26, 34]. The data collected were employed to investigate the surface chloride content and the chloride diffusion coefficient in terms of concrete mixtures, exposure conditions and time [4]. The

**Fig. 1.1** Marine atmospheric exposure site in Mexican gulf, Port Progresso, Mexico. The exposure site was designed to obtain atmospheric results of chloride deposition, soluble and insoluble solids, sulfates, carbon dioxide, temperature, relative humidity and pluvial precipitation. Temperature is in the range of 26–40 °C with temperatures surrounding 70–95%. This exposure site is being used to check durability of several types of specimens with different cements, dosage, inhibitors etc. *Source* Courtesy of Dr. Pedro Castro-Borges

first sites of exposure in Latin America were set up in Mexico in 1991 [6], cf. Fig. 1.1. Through the DURACON project, the chloride ingress and carbonation were investigated for concrete specimens exposed in the Ibero-American urban, marine and rural environments [33]. In Cuba, the Punta Matamoros site was established in 2015 in Cayo Santa María on the shore of Caribbean Sea to observe the real performance of concretes using LC3 binders [20], cf. Fig. 1.2.

### *1.2.2 Exposure Sites in Europe*

The Träslövsläge site was established in Sweden in 1990 to investigate the chloride penetration in concrete materials and the relevant influencing factors [27], cf. Fig. 1.3. The obtained results helped the calibration of ClinCon model and the service life prediction of concrete structures with respect to chloride ingress [31]. A French marine exposure site in La Rochelle dates back to 1920s, cf. Fig. 1.4. The systematic observations starting in 1996 in the tidal zone of the site helped to quantify the performance of concretes with different binders in real marine environments [5]. Another study on the same site starting in 2005 aimed to monitor corrosion initiation in reinforced concrete with different water-to-cement ratios [29]. At the Huelva site,

**Fig. 1.2** The Punta Matamoros exposure site is within 500 m from the seashore in Cayo Santa Maria, Cuba. The average temperature and humidity are respectively about 28 °C and 80%, with regular strong winds (Hurricane). This site has been mainly used for concretes containing LC3 cementitious materials. *Source* Courtesy of Prof. José Fernando Martirena-Hernández

a marine beach on the southwest coast of Spain, four concrete blocks (50 × 50 × 200 cm in size illustrated in Fig. 1.5) were exposed in 1989, to study the resistance to seawater attack of different cement types. The investigations were further extended to study the reinforcement corrosion [1]. In parallel, small prism specimens (20 × 10 × 15 cm in size) containing steel reinforcement with different concrete covers have been immersed in a large tank with facilities to continuously pump natural seawater from the Mediterranean Sea in CSIC, Barcelona since 1982, to study the chloride ingress [2].

The Cabo Raso exposure site, established in 2002 on the west coast of Portugal, was designed to collect atmospheric data on chloride and sulfate depositions, $CO_2$ concentration, temperature, relative humidity, pluvial precipitation, and time of wetness. The site is located adjacent to a weather station, providing comprehensive environmental data, cf. Fig. 1.6. The exposure site features three distinct zones with concrete and reinforced concrete specimens of various compositions: an air-exposed zone, a tidal zone, and a submerged zone, with the latter two situated in a decommissioned shellfish farm. The obtained data also contributed to the DURACON project, which examined the impact of urban and marine meteoro-chemical parameters on the performance of reinforced concrete structures in nine Ibero-American countries [33].

The Rødbyhavn marine exposure site was established in the western part of the Baltic Sea, Denmark, in 2010 by Femern A/S, as part of the preparatory work for the Fehmarn Belt Fixed Link between Denmark and Germany. Concrete panels of 2 m × 1 m × 0.2 m are placed vertically with an average sea level 0.7 m from top

Fig. 1.3 The Träslövsläge field site was built in 1990 on the west coast of Sweden. The local temperature of seawater is 11 °C and the chlorides concentration is 10–18 g/L. More than 40 types of concretes with different binders were exposed. The results are used for chloride diffusivity regression and model validation. Courtesy of Prof. Luping Tang

Fig. 1.4 Tidal zone in the immediate vicinity of the Tour Saint-Nicolas in La Rochelle city, on the west coast of France. Concrete and reinforced concrete blocks of various sizes and compositions are exposed to marine tidal cycles for studies on chloride diffusivity and reinforcement corrosion. *Source* Courtesy of Dr. Rachid Cherif

[12], cf. Fig. 1.7. Freezing temperatures are moderate and relatively rare and the sea levels inside the harbour allow to cover submerged, tidal and splashing zones for the panels. The exposure site was moved to Spodsbjerg harbor 50 km away in 2024, maintaining the concrete panels at the same seawater level. Specimens of more than 15 different concretes are installed with sensors to monitor corrosion initiation and determine chloride threshold values [30]. Along the coast of north Adriatic coast,

**Fig. 1.5** Reinforced concrete blocks 50 × 50 × 200 cm in size, semi-buried at two distances from the lower sea border near the frontier of Portugal at the Atlantic Ocean. *Source* Courtesy of Dr. Carmen Andrade

**Fig. 1.6** The atmospheric zone (left) and tidal zone (right) of Cabo Raso exposure site, on the west coast of Portugal. The site experiences an average air temperature of 14 ºC and relative humidity of 77%, with regular onshore winds. The average chloride concentration in the sea water is about 18–21 g/L, and in the air, the deposition rate is approximately 1215 mg $m^{-2}$ $day^{-1}$. *Source* Courtesy of Dr. André Valente Monteiro

an exposure site was established in 2009, as a part of European project ARCHES, on the launch ramp of Krk bridge to monitor the corrosion of embedded reinforced steel of different types in 18 concrete columns, providing sound basis for retrofit of durability deterioration of this bridge [28].

### 1.2.3 Exposure Sites in Asia and Africa

Japan has started the marine exposure practice since 1980 [22, 32], and the obtained results were used to study the durability mechanisms of reinforced concretes with/without coatings [23, 24]. The first exposure stations in China were built in marine ports in 1980s on the southeast coast of China mainland [35]. The observations

**Fig. 1.7** Concrete panels (2.0 × 1.0 × 0.2 m) exposed at the Rødbyhavn harbor field exposure site, built in 2010 on the south coast of Lolland in Denmark and moved to Spodsbjerg harbor at the east coast of Langeland in 2024. The local average sea water temperature is 9 °C and the average chloride concentration is 7 g/L. More than 15 concrete types with different binders are exposed. *Source* Courtesy of Dr. Henrik Erndahl Sørensen

and results have supported the drafting and revision of several Chinese technical standards for concrete durability [9, 10]. The exposure data from Zhanjiang station, cf. Fig. 1.8, were used to calibrate the durability model in the design works of the Hong Kong-Zhuhai-Macau sea link project [17].

Systematic studies in exposure conditions in South Africa have been undertaken since the 1990s at various marine exposure sites around Cape Town and Durban, cf. Fig. 1.9. The studies aimed at quantifying the processes of chloride ingress, rebar corrosion, and penetrability characteristics of different types of concretes [13, 19, 25]. The collected data were used to investigate the impact of such factors as alternative binders, mix proportions, curing practices and crack width on the concrete performance under different exposure conditions, and to calibrate service life prediction models.

## 1.3 Marine Exposure Zones

In marine environment, the mechanisms and intensity of seawater actions on the concrete depend much on the interaction between the seawater, the aerodynamics of marine air and the concrete surface. The detailed mechanisms and influential factors will be given in Chap. 2, for which the exposure zone is the very pertinent concept. The definition of exposure zones is essential for identifying the intensity of environmental

**Fig. 1.8** Zhanjiang exposure station is situated on the South-eastern coast of China. The exposure station was established in 1987, located in the wharf of Zhanjiang port (average temperature 22.5 °C and relative humidity 85%), and has accumulated data on chloride ingress for more than 30 years. *Source* Courtesy of Mr. Jianbo Xiong

actions and also the design of exposure stations. These exposure zones are described by the local marine climate created by the exposure sites (stations) for the specific placement of specimens, which are generally defined through the seawater hydrology parameters. The definition of these exposure zones [16] is given below and will be used throughout this report.

The SUB zone: The submerged zone refers to the condition permanently exposed to seawater. The exposure environment follows mainly the characteristics of the seawater, including its temperature and chemical composition. Also, marine biological growth can occur on the surface of concrete structures and specimens. The upper limit for SUB zone is related to the lowest astronomical tide level and wave height [11] for unshielded marine environment, and to the design low water level for shielded marine environment [8].

The TID zone: The tidal zone refers to the condition of concrete subject to alternating immersion by tidal water and atmospheric exposure to marine air. The seawater tide follows approximately 12-h cycles. The exposure environment in TID zone relates to the characteristics of both seawater and marine air. Accordingly, the durability processes include both seawater-induced ones and marine air induced ones, and surface biological growth can also occur. The lower limit of TID zone is the upper limit of SUB zone. The extended range (upper limit) of TID zone considers mainly the highest astronomical tide and effective wave height for unshielded marine

**Fig. 1.9** The Granger Bay exposure site is situated in Cape Town, South Africa. The average seawater temperature at this site is 12–15 °C and the strong winds cause heavy wave actions, which produces quite aggressive exposure conditions. The obtained data are mainly used for model calibration. This image was originally published in [13] and is licensed under a Creative-Commons license CCBY 4.0

environment, and the design high water level for shielded marine environment [8]. In some cases, the TID zone is combined into the SPL zone.

The SPL zone: The splash zone refers to the exposure of concrete surface to both the splashing actions of seawater waves and the marine air. The splashing actions are created by the collision between the seawater waves and the concrete surface, and thus they are highly dependent on the local arrangement of exposure structures and specimens. The concrete surface in SPL zone will be subject to actions from both the seawater splashing and the marine air. The chloride accumulation and the subsequent steel corrosion are substantially promoted by the drying-wetting cycles and the alternative availability of moisture and oxygen in air [7]. For unshielded marine environments, the upper limit of SPL zone is related to the highest astronomical tide and wave height [11], and to design high water level for shielded marine environment [8]. As the exposure site is arranged into a shielded structure, the splashing actions will be specific to the arrangement of the exposure spaces, and the appropriate placing of concrete specimens should resort to the local observation and recordings of the water movement.

The ATM zone: The atmospheric zone refers to the exposure of concrete surface to marine air only. The exposure environment of concrete surface is the marine atmosphere, and the main environmental actions include the temperature, the airborne chlorides, the $CO_2$ and the natural precipitation. The lower limit of ATM zone is the upper limit of SPL zone, and the intensity of the airborne chlorides depends much on the local aerodynamics and the height/distance from the sea level and the coastline [18]. For concrete specimens exposed in ATM zone, the chloride deposition rate is one of the environmental factors to be recorded.

## 1.4 Technical Committee of TC 289-DCM

The TC work was divided into three work groups (WG): (1) the establishment of long-term exposure database for structural concrete exposed to natural marine environments, and the correct presentation of long-term exposure data; (2) the exploitation of the long-term exposure and observation data via mechanism interpretation, and the rational indicators for engineering use; (3) the application of long-term exposure/observation data and their exploitation to the service life design and management of concrete infrastructures. The technical contents of these WG works are detailed in the following.

### *1.4.1 WG1: Long-Term Exposure Database and the Correct Presentation*

On the basis of the long-term exposure (field) stations accessible to the TC members, the first aim of this work phase was to establish a long-term exposure and observation database of structural concretes. Since the TC members had extensive geographic coverage, this database bears more global sense compared to the databases of same type. As the first step of this operation, the focus was put on the natural marine exposure conditions, and both the long-term data from exposed concrete specimens and extracted cores were involved. The collection of long-term data covered also the concretes with protection measures (silane impregnation and coatings), and concretes with cracks. Then, these data were arranged following a standard procedure, including the original concrete proportioning, concrete raw materials, exposure conditions (thermal, moisture and aggressive agents), and the deterioration performance (penetration depths, profiles, visual state, etc.), from which a normalized format of exposure data was generated. The objective of this phase, apart from the exposure database, was to propose a standard format for complete description of long-term exposure data, which serves as the basis for better understanding and interpretation of the long-term durability performance of structural concretes.

### 1.4.2 WG2: Rational Interpretation of Long-Term Exposure/ Observation Data

So far, the most widely used parameter to represent the long-term durability performance is the apparent chloride diffusivity, and several RILEM TCs have worked on this subject. This work phase differed from the apparent chloride diffusivity study and focused on the correct mechanism interpretation of the long-term data. Actually, in exposure conditions, several fundamental processes exist simultaneously: the ingress of marine chlorides, the ingress of other aqueous ions (sulfates), the carbonation of surface layer (in air-exposed zones), the leaching (in water immersion/splashing zones) and/or frost actions on the concrete surface (tidal zones in cold climates). Using one apparent chloride diffusivity is much too simplistic to obtain meaningful results for engineering application. Accordingly, this phase of work started with a review of available models of chloride ingress in marine environments, followed by a critical analysis on their capacity of prediction via the review on the model benchmarking. On this basis, new engineering indicators were proposed and exploited, providing deeper insight into the regressed "apparent chloride diffusivity". A similitude issue involved the behaviors of same concretes exposed at different sites: the understanding of the difference will help the prediction of concrete durability performance at one site different from the exposure site, which would greatly reduce the cost of multiple exposure stations. Moreover, the dispersion of long-term exposure among specimens was subject to specific study, quantifying the statistical characteristics related to the exposure conditions and the performance. Overall, the second work phase was to address better understanding and better interpretation of long-term data, via new engineering indicators in addition to the apparent chloride diffusivity.

### 1.4.3 WG3: Technical Requirements for Long-Term Field Exposure Stations and the Application

So far, the implementation of long-term exposure data into the service life design and management of concrete infrastructures are often user-tailored, and far from systematic. This work phase, on the basis of the correct presentation of long-term exposure data and the rational interpretation of these data, established technical requirements for long-term field exposure stations in views of more efficient and sustainable exploitation of long-term exposure data. These requirements reflected the state-of-the-art knowledge accumulated on the subject, and served as the technical guideline for new exposure stations construction and upgrading of existing stations. Secondly, this phase continued on the rational application of long-term data to the service life design and management of concrete infrastructures, which was split into two topics: how to help the durability quality control in new constructions, and how to help the durability planning during the service life phase. To this purpose, several engineering applications, including several large-scale projects in marine environments, were reviewed for their exploitation of long-term exposure data for service life management. Overall, this phase of work provided the basic technical requirements for long-term exposure stations, and the real cases for the integration of long-term exposure data into the service life design and management of concrete infrastructures.

The above works have been performed during the period of 2019–2024, attracting a total of 80 members in the TC representing 26 countries and regions from six continents. From the works, TC published two publications in the journal of Materials and Structures [15, 16]. Moreover, a first world-wide database has been established for structural concretes in marine exposure conditions, hosting more than 2000 sets of data. The database is presently hosted on local servers and expected to be accessible to wider scope in near future.

## 1.5 Structure of STAR Report

The STAR report is divided into seven chapters. Chapter 2 describes the durability mechanisms involved in marine exposure for structural concretes, and the related influential factors. Chapter 3 focuses on the presentation of the data generated from the long-term exposure practice, including also the sampling and testing, and the dispersion characteristics of exposure data. The world-wide database developed in TC 289-DCM is also presented in this chapter for their basic functions and accessibility. Chapter 4 is dedicated to the interpretation and exploitation of the obtained exposure data, touching the modeling aspects and engineering indicators. The guideline for designing and operating exposure sites is provided in Chap. 5, on the basis of the available experiences from major exposure stations. Chapter 6 collects several engineering cases using the long-term exposure practice as major support for service life and maintenance planning. General conclusions and perspectives are given in Chap. 7.

## References

1. Andrade, C., Sagrera, J.L., Sanyuán, M.A.: Several years study on chloride ion penetration into concrete exposed to Atlantic Ocean water. In: Testing and Modelling Chloride Ingress into Concrete – Proceedings of the 2nd International RILEM Workshop, RILEM PRO 19, Paris, France, pp. 121–134 (2000)
2. Andrade, C.: Steel corrosion rates in concrete in contact to sea water. Cem. Concr. Res. **165**, 107085 (2023)
3. Alexander, M.G.: Marine Concrete Structures, Design, Durability and Performance. Woodhead Publishing, Amsterdam (2016)
4. Alexander, M., Thomas, M.: Service life prediction and performance testing - current developments and practical applications. Cem. Concr. Res. **78**, 155–164 (2015)

5. Baroghel-Bouny, V., Dierkens, M., Wang, X., Soive, A., Saillio, M., Thiery, M., Thauvin, B.: Ageing and durability of concrete in lab and in field conditions: investigation of chloride penetration. J. Sustain. Cem. Based Mater. **2**(2), 67–110 (2013)
6. Castro-Borges, P., Maldonado, L.: Initial efforts to evaluate the corrosion problems in the infrastructure of the Mexican southeast coastal zone. In: CORROSION/95, Annual Congress of NACE-International, Paper No. 21, 1995, Orlando, Florida (1995)
7. CEB: Durable Concrete Structures – CEB Design Guide, Bulletin No. 183, Comité Euro-International du Béton, Lausanne (1992)
8. CIS: Corrosion prevention technical specifications for concrete structures of marine harbors engineering (JTJ 275–2000), China Industrial Standard. China Transportation Publishing, Beijing (2000)
9. CIS: Standard for Durability Design of Port and Waterway Engineering Structures (JTS 153–2015), China Industrial Standard. China Transportation Publishing, Beijing (2015)
10. CNS: Standard for Design of Concrete Structure Durability (GB/T 50476–2019), China National Standard. China Architecture Publishing, Beijing (2019)
11. DNV: Offshore standard DNV-OS-C101: design of offshore steel structures, general (LRFD method), Det Norske Veritas, Høvik, Norway (2011)
12. Fjendbo, S., Sørensen, H.E., De Weerdt, K., Jakobsen, U.H., Geiker, M.R.: Correlating the development of chloride profiles and microstructural changes in marine concrete up to ten years. Cem. Concr. Compos. **131**, 104590 (2022)
13. Heiyantuduwa-Beushausen, R.: The influence of concrete mix composition and environmental exposure on long-term chloride ingress in concrete. Ph.D. thesis, University of Cape Town, South Africa (2022)
14. Li, K.F.: Durability Design of Concrete Structure: Phenomena, Modelling and Practice. Wiley, London (2016)
15. Li, K.F., Zeng, J.J., Tang, L.P., Sorensen, H.E., Castro-Borges, P. et al.: Long-term field exposure of structural concretes in marine environment: state-of-the-art review by RILEM TC 289-DCM. Mater. Struct. **55**, 205 (2022)
16. Li, K.F., Zeng, J.J., Tang, L.P., Sorensen, H.E., Andrade, C., et al.: RILEM recommendation from TC 289-DCM: guideline for designing and operating long-term marine exposure sites. Mater. Struct. **57**, 44 (2024)
17. Li, K.F., Zhang, D.D., Li, Q.W., Fan, Z.H.: Durability for concrete structures in marine environments of HZM project: design, assessment and beyond. Cem. Concr. Res. **115**, 545–558 (2019)
18. Lindvall, A.: Environmental actions on concrete exposed in marine and road environments and its response-consequences for the initiation of chloride induced reinforcement corrosion. Ph.D. thesis, Chalmers University of Technology, Sweden (2003)
19. Mackechnie, J.R.: Predictions of reinforced concrete durability in the marine environment. Ph.D. thesis, University of Cape Town, South Africa (1996)
20. Martirena-Hernández, F., Scrivener, K.: Introducing a novel cement. Int. Cem. Rev. **9**, 1–3 (2019)
21. Moffatt, E., Thomas, M.: Performance of 25-year-old silica fume and fly ash lightweight concrete blocks in a harsh marine environment. Cem. Concr. Res. **113**, 65–73 (2018)
22. Mohammed, T., Yamaji, T., Aoyama, T., Hamada, H.: Marine durability of 15-year-old uncracked and precracked concrete specimens made with different cements. Proc. JSCE **697**(V-54), 295–309 (2002)
23. Mohammed, T., Hamada, H., Yamaji, T.: Concrete after 30 years of exposure - part I: mineralogy, microstructures, and interfaces. ACI Mater. J. **101**, 3–12 (2004)
24. Mohammed, T., Hamada, H., Yamaji, T.: Concrete after 30 years of exposure - part II: chloride ingress and corrosion of steel bars. ACI Mater. J. **101**, 13–18 (2004)
25. Otieno, M., Beushausen, H., Alexander, M.: Chloride-induced corrosion of steel in cracked concrete - part I: experimental studies under accelerated and natural marine environments. Cem. Concr. Res. **79**, 373–385 (2016)

26. Roshore, E.: Cement durability program, long-term field exposure of concrete columns. US Army Engineer Waterways Experiment Station, Vicksburg (1972)
27. Sandberg, P., Tang, L.P., Andersen, A.: Recurrent studies of chloride ingress in uncracked marine concrete at various exposure times and elevations. Cem. Concr. Res. **28**, 1489–1503 (1998)
28. Serdar, M., Bjegović, D.: Performance of concrete reinforced with corrosion resistant steels exposed to real marine environment. In: International Conference on Materials, Corrosion, Heat treatment, testing and tribology, Zadar, Croatia, 4–7 October 2017 (2017)
29. Sleiman, H., Amiri, O., Aït-Mokhtar, A., Loche, J.M, Tahlaiti, M.: Towards the prediction of service life of concrete structure in unsaturated saline environment. In: 13th International Congress on Structural Faults and Repair, Edinburgh, Scotland, 15–17 June 2010 (2010)
30. Sørensen, H.E., Poulsen, S.L., Jönsson, U.: Chloride threshold values from concrete blocks exposed at Rødbyhavn field exposure site. In: Proceedings of FIB Symposium Performance-based Approaches for Concrete Structures, Cape Town, South Africa, 21–23 November 2016, 10pp. (2016)
31. Tang, L.P.: Engineering expression of the ClinConc model for prediction of free and total chloride ingress in submerged marine concrete. Cem. Concr. Res. **38**, 1092–1097 (2008)
32. Toshio, H., Kento, U.: Permeation of chloride into carbonated concrete exposed to marine environment. Concr. Eng. Annu. Proc. **17**, 847–852 (1995). (In Japanese)
33. Troconis de Rincón, O., Montenegro, J.C., Vera, R., et al.: Reinforced concrete durability in marine environments DURACON project: long-term exposure. Corrosion **72**(6), 824–833 (2016)
34. Uno, J., Robertson, I., Newtson, C.: Corrosion susceptibility of concrete exposed to a marine environment. Research report UHM/CEE/04-09, Department of Transportation Federal Highway Administration, University of Hawaii, US (2004)
35. Wang, S., Tian, J., Fan, Z.H.: Research on theory and method of service life prediction of marine concrete structures based on exposure test and filed investigation. China Harb. Eng. **169**, 67–74 (2010). (in Chinese)

# Chapter 2
# Concrete in Marine Environments: Mechanisms and Factors

**Mette Rica Geiker, Kefei Li, Junjie Zeng, Xuemei Liu, Pedro Castro Borges, Abdelkarim Aït-Mokhtar, Rachid Cherif, Fragkoulis Kanavaris, Gilberto Cidreira Keserle, Riccardo Maddalena, José Fernando Martirena Hernandez, and Carmen Andrade**

M. R. Geiker (✉)
Department of Structural Engineering, Norwegian University of Science and Technology, Trondheim, Norway
e-mail: mette.geiker@ntnu.no

K. Li
Civil Engineering Department, Tsinghua University, Beijing, China

J. Zeng
CCCC Fourth Harbor Engineering Co., Ltd., Guangzhou, China

X. Liu
Department of Infrastructure Engineering, The University of Melbourne, Parkville, Australia

P. C. Borges
Department Applied Physics, Cinvestav-Mérida, Mérida, Yucatán, México

A. Aït-Mokhtar · R. Cherif
LaSIE UMR CNRS 7356, La Rochelle University, La Rochelle, France

F. Kanavaris
Specialist Technology Analytics Research, Materials, Arup Group Ltd, London, UK

G. C. Keserle
Department of Expertise, Englobe Corp, Québec, Canada

R. Maddalena
School of Engineering, Cardiff University, Cardiff, UK

J. F. M. Hernandez
Facultad de Construcciones, Universidad Central de las Villas, Santa Clara, Villa Clara, Cuba

C. Andrade
International Center for Numerical Methods in Engineering (CIMNE), Barcelona, Spain

 
K. Li et al. (eds.), *Long-Term Performance of Structural Concretes in Marine Exposure Conditions*, RILEM State-of-the-Art Reports Volume,
https://doi.org/10.1007/978-3-032-06959-7_2

## 2.1 Deterioration Mechanisms of Concrete in Marine Environment

Concrete is a heterogeneous composite material consisting of aggregates embedded in cement paste. Due to geometrical restrictions, the distribution of cement particles and hydrates is affected by the aggregates creating an interfacial transition zone (ITZ) between the aggregates and the unaffected bulk paste; similar zones are observed at reinforcement and other embedded objects. Cement paste is initially a suspension of cement grains in water. Gradually the cement[1] reacts with water (hydrates) and the cement paste becomes a hardened porous composite. Depending on the initial water-to-cement ratio (w/c), and the type and degree of reaction of the cement, the pore structure and thus the transport rate within the cement paste vary. In addition to the pore structure, the rate of mass transport is affected by the degree of saturation and the possible interaction between the pore solution and the solids.

Exposed in marine environments, structural concrete will undergo physical and chemical changes due to both the environmental actions (see Fig. 2.1) and the service loads [78]. These changes depend highly on the local exposure conditions. For submerged concrete, i.e. elements under sea level, the pore liquid is permanently in connection with the seawater, resulting in the diffusion of ions, such as $Cl^-$, $Mg^{2+}$, $SO_4{}^{2-}$, from the seawater into the concrete and the outward transport (leaching) of ions, such as $Ca^{2+}$ and $OH^-$, from the pore solution to the seawater. In addition, transport of ions might take place in form of convection of seawater by capillary absorption (suction) into non-saturated concrete and/or by permeation due to hydraulic pressure. All these transport processes will alter the liquid–solid equilibrium between the pore solution and the solid phases in the matrix of the concrete [63]. For atmospheric exposed concrete, air-borne aerosols of seawater can deposit on the surface and ions ingress primarily by diffusion. Due to a lower moisture content of atmospherically exposed concrete and typically a lower surface concentration of ions, the diffusion process is less active here compared to submerged concrete [114], but the $CO_2$ in the atmosphere will induce carbonation on the concrete surface potentially changing the chloride binding capacity. In case of de-icing of traffic lanes, exposure to salt spray may also take place in the ATM zone [32]. For the TID and SPL zones, the involved processes are intermediate to the above two cases, except that drying-wetting cycles will cause the transport to be dominated by capillary absorption (suction) in the outermost concrete layer, creating a so-called convection zone [28]. In addition, the direct contact with seawater can induce pure mechanical damage of concrete surface by erosion (seawater flow) or by abrasion (sediments or/and flowing ice) [146], under cold climates, freeze–thaw damage may occur in highly saturated concrete as well [33].

In the following the typical deterioration mechanisms acting on marine reinforced concrete (RC) structures are described covering physical (freeze–thaw action and ice abrasion), chemical (seawater action and carbonation), electrochemical (corrosion),

---

[1] Cement is here used as a general term for binders with cementitious properties.

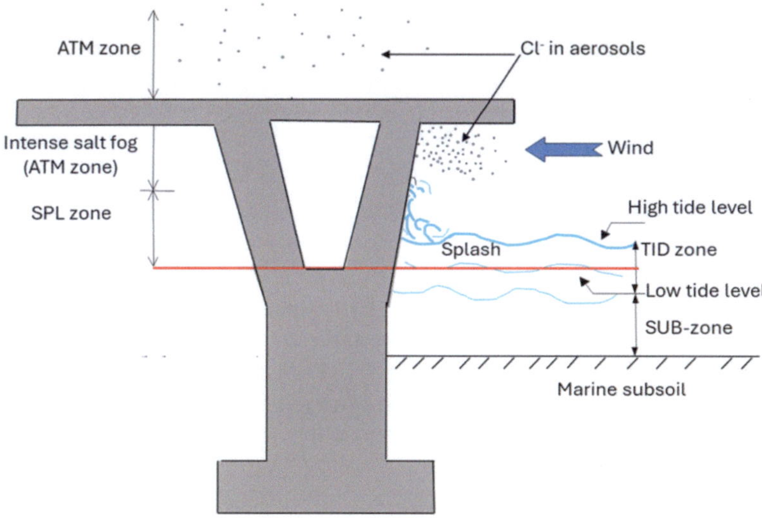


**Fig. 2.1** Schematic illustration of the environmental loads and the four exposure zones typically used for marine structures: SUB, TID, SPL and ATM. Potential deicing of traffic lanes might affect part of the concrete in the ATM zone

and biological processes. So far, steel corrosion is the major durability concern for RC structures in marine environment. The remaining processes may alter the concrete properties and potentially disturb the electrochemical stability of imbedded steel bars [99, 145]. The following sections are arranged from the ATM zone to the zones with direct seawater contact.

### *2.1.1 Carbonation*

In marine ATM and SPL zones, the concrete surface is exposed to carbon dioxide ($CO_2$), cf. Fig. 2.1. Depending on the moisture content of concrete surface, the $CO_2$ present in the atmosphere diffuses into the concrete through the pore structure and causes the carbonation of the cement hydrates. The direct consequence of carbonation is the consumption of hydrates such as portlandite ($Ca(OH)_2$ or CH), and calcium silicate hydrates (C-S-H), and formation of $CaCO_3$, decreasing the pH value of the pore solution.

In marine exposure, the carbonation in the ATM and SPL zones is regarded as a concurrent process along with the chloride accumulation on concrete surface and ingress into the bulk part of concrete. The carbonation alters both the physical and chemical properties of concretes. Physically, the $CaCO_3$ precipitation and the free water released from the carbonation reactions will change the microstructure and the degree of pore saturation [113, 125]. These studies on hardened cement pastes and mortars showed a decrease of total porosity but increase of capillary porosity (>30 nm) by carbonation. Chemically, the carbonation decreases the pH value of pore solution to around 9.0 [118], and the chemically bound chlorides in Friedel's salts will be released into aqueous phases. Also, the carbonation of C-S-H, main physical adsorbent of chlorides, will contribute to the chloride release [126]. The combined decrease of the pH value of pore solution will cause an increase in the local $Cl^-$/ $OH^-$ ratio and promote the corrosion of steel.

As carbonation is highly dependent on the moisture content of concrete, the carbonation extent highly varies between structures. Laboratory studies have shown an overall acceleration of chloride ingress by carbonation [16, 88, 153]. Due to the increased diffusion of carbon dioxide in concrete at a lower degree of saturation, the ATM zone experiences more significant carbonation compared to the other exposure zone. In Scandinavia, the high degree of saturation and typical low w/c of marine concretes effectively limit the carbonation extent.

### *2.1.2 Seawater Action on Concrete*

As a general feature, it has been shown that exposure to seawater causes changes to the outermost concrete layer. Collating data from field exposed concrete in Denmark and Norway, Jakobsen et al. [67] found a binder independent sequence of elemental zonation in submerged concrete starting with a thin Mg-rich zone at the surface, followed by a S-rich zone, and subsequently a Cl-rich zone, in addition, formation of a calcite/brucite crust on the surface was typically observed. The feature of elemental zonation is illustrated in Fig. 2.2. Fjendbo et al. [44] showed that the elemental zonation develops gradually from the exposed surface and found that the depth of microstructural changes correlated with the position of the maximum chloride content (chloride peak). Investigating trial mixes for the Femern Belt connection between Denmark and Germany, the depth of the microstructural changes was 4–10 mm after ten years of exposure, smallest in a slag containing concrete [44].

### *2.1.3 Chloride Ingress and Binding*

During the process of ionic transport in concrete, chlorides undergo chemical interactions with anhydrous compounds present in cement [6, 115, 138, 145]. Additionally, chlorides are subject to adsorption by the C-S-H, involving both electrochemical interactions and the formation of an electrical double layer [5, 19, 144]. The chemical interactions result in the formation of compounds that precipitate within the pore solution. Sutter et al. [138] emphasized the dissolution of portlandite during chloride transfer, leading to the formation of Friedel's salt and hydrated calcium oxychloride ($3CaO\ CaCl_2\ 15H_2O$). These occurrences significantly impact the microstructure of concretes, pore solution and their chloride diffusion coefficients.

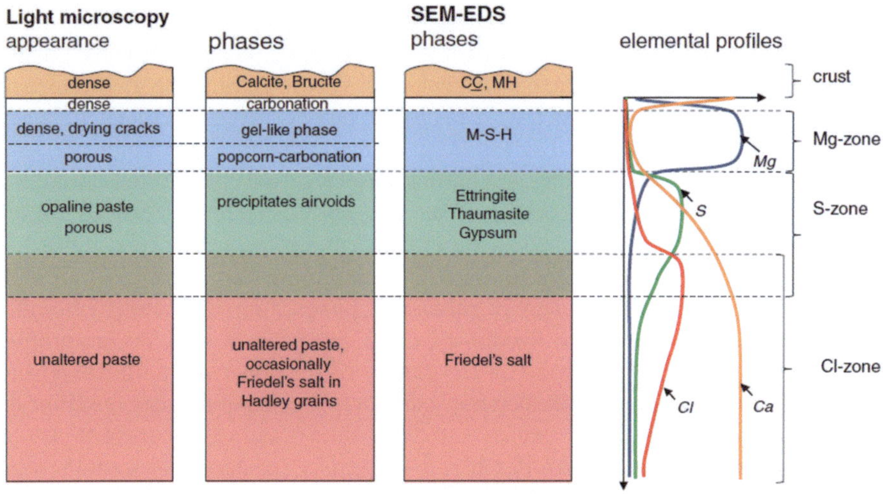


**Fig. 2.2** Features of elemental zonation in the outer surface of marine submerged exposed concrete. Image redistributed from [67] and licensed under CC BY-NC-ND 4.0 (https://creativecommons.org/licenses/by-nc-nd/4.0/)

Midgley and Illston [103], and Kayyali and Haque [72] observed changes in the porosity of Portland cement pastes after 28 and 90 days of immersion in sodium chloride solutions (with concentrations of 30 and 39 g/l, respectively). The findings highlighted that chloride diffusion reduced and refined the porosity by promoting the formation of calcium chloride on the surface of C-S-H. Regourd et al. [122] noted that chlorides altered the initially fibrous C-S-H structure to a reticulated morphology. Moreover, [74] analyzed the pore size distribution of mortars partially submerged in a 7% NaCl solution, revealing a decrease in total porosity and refinement of the pore network due to chloride diffusion through the fully immersed part of the material. Sutter et al. [138] reported the formation of calcium oxychloride within the microstructure of mortar immersed in a 15% $MgCl_2$ solution. This alteration of the microstructure impacts the chloride diffusion coefficient of the material.

Regarding the chemical composition of the pore solution, the presence of external ions like chlorides leads to a thermodynamic imbalance in the hydrates—pore solution system, facilitating the Portlandite dissolution [29, 31, 51] and the precipitation of compounds such as Friedel's and Kuzel's salts [100, 119, 137]. Buckley et al. [20] and Kyle et al. [76] conducted an extraction and analysis of pore solutions from various mortars and cement pastes, all of which contained sodium chloride in the mixing water. In the case of 7.5% chloride by mass of cement, the free chloride content in the pore solution peaked at approximately 16%, which is presumed to signify the saturation point of the solution, with a pH about 13.5.

The submerged concrete, chloride ingress is primarily coinciding with leaching of hydroxyl ions. As part of the phase changes taking place in marine exposed concrete, calcium hydroxide leaching has been observed [148]. Both De Weerdt et al. [148]

and later Fjendbo et al. [44] observed calcium hydroxide depletion to the same depth as the chloride peak in plain PC and PC + SF concrete, but to relatively greater depths for FA containing concrete. Machner et al. [91] illustrated the importance of leaching on the chloride ingress rate, the chloride ingress depth was much higher in a specimen exposed to a neutral salt solution than a salt solution with high pH.

In concrete exposed to wetting and drying, two zones are typically observed: the convection zone and the bulk concrete. The convection zone is a superficial porous zone that dries and wets due to interaction with the environment. This zone is affected by both chloride ingress and calcium hydroxide leaching. The internal denser bulk concrete experiences chloride ingress, but there is no leaching of calcium hydroxide. At the interface between these two zones, there is a peak in the chloride profile, the position of which will depend on the quality of the concrete and the exposure conditions. The distance from the concrete surface to the position where the chloride peak occurs is variable [23].

Examples of total chloride profiles after 8–22 years' exposure in the south-eastern part of Denmark are given in Fig. 2.3. The profiles were obtained by profile grinding of extracted cores, dissolution of powder with acid and subsequent chloride analysis. The figure illustrates the impact of exposure conditions (SUB and SPL zone at different levels) and exposure time (8–22 years). Note that profiles from in principle identical cores vary with respect to both time and level. Note also that a peaking behavior is not only observed for in the SPL zone where a convection zone is expected, but also in the SUB zone. This so-called peaking behavior, where the maximum chloride content with time increases in both value and depth can be explained by the gradual progress of leaching for submerged concrete. Hemstad et al. [59] documented the impact of pH on the binding capacity of chlorides, with an increased binding at moderate pH reduction followed by the loss of chloride binding at lower pH.

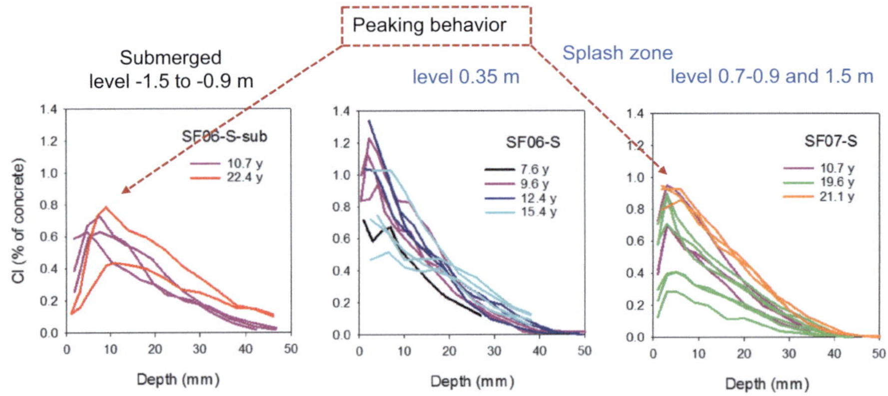


**Fig. 2.3** Total chloride profiles from piers of the Farø bridges, Denmark after 8–22 years' exposure illustrating the impact of exposure conditions (SPL and SUB) and exposure time. Peaking behaviour explained by microstructural changes (SUB zone) and convection (SPL zone). Individual figures from [49]

The observation of a gradually developing microstructural changes and their impact on the chloride profile calls for reconsideration of the assessment method of the residual service life of marine exposed structures. In the case of obtaining chloride ingress data from subsequent inspections in time scale, the so-called square-root method seems promising [43, 116]. This method is based on the observations of a square-root time law for chloride ingress rate in the non-leached bulk concrete.

### 2.1.4 Ice Abrasion

In arctic sea regions, concrete structures are often subjected to heavy mechanical loads near the water level due to the drifting ice sheet and the impact that the ice sheet generates. Such abrasion can cause loss of thickness and strength of the concrete structural members which have to bear significant ice pressure [64, 131]. Damages of concrete structures caused by ice abrasion have often been observed in bridge piers and offshore structures like lighthouses and oil-platforms in arctic and subarctic sea waters and sea windmill foundations [65]. The erosion of such structures, in most cases, is caused not only by sea ice abrasion but also by freezing and thawing [64] and does threaten the structural integrity and functionality of structures throughout their service life. However, it is difficult to separate abrasion-related damage from freezing–thawing related damage in field observations. The abrasion depth can progress for several centimeters over a structural lifetime, deteriorate the inner concrete layers away and abrade the reinforcement. It is understood that the wearing rate depends mainly on ice temperature and contact pressure. For abrasion, relative velocity, concrete aggregate type, and compressive strength of concrete are reported to be not as critical as ice temperature and contact pressure [64]. The surface treatment such as polyurethane resin and resin mortar linings, which reduce the friction between structures and sea ice, was found to be very effective for wear protection [64]. When designing, applying and maintaining the surface treatments, one needs to consider the risk of freeze–thaw damage due to potential accumulation of water causing the degree of saturation of the concrete to surpass a critical value.

Shamsutdinova et al. [134] exposed structural concrete surfaces to sliding freshwater ice under 1 MPa pressure for 3 km of sliding distance. They found a severe-to-mild wear transition with the maximum wear rate in the first sliding kilometer. The coefficient of friction was high when ice consumption was high due to ice spallation and pulverization, whereas the wear was not directly correlated to the coefficient of friction. The abrasive wear or consumption of the ice was in the order of 30,000–100,000 times that of concrete despite of its strength and stiffness being only 1–10 times lower than that of concrete. A five-year data collection on two ice shield piers of the Confederation Bridge was reported in 2007 and showed an average abrasion loss of 0.31 mm/year, removing both paste an aggregate from the piers [131].

The collision of an ice sheet with a concrete structure causes various kinds of interaction. The structure's geometry, the ice load conditions, and ice properties affect the nature and magnitude of the global load [90, 134]. The ice abrasion damage of concrete could be attributed to three distinct mechanisms: (1) asperities of ice sliding on concrete can create tensile stress sufficient to initiate cracks with fatigue from repeated contacts playing an important role; (2) debris released from the surface of the concrete can act as sharp abrasive particles that enhance the damage as the ice drags them along the surface; and (3) water forced into surface defects can create pressure that propagates cracks [65].

A typical interaction between an ice sheet and a concrete column was described by Jacobsen et al. [65]. If in direct contact, high normal stresses are imposed on both ice and concrete, and an intermediate layer of water is forced against both surfaces. The normal forces diminish, and shear forces rise in the region, where debris are dragged along the surface of the concrete. This is the zone where the greatest abrasion damage is found to occur. In the further away region, the normal force becomes smaller, but the quantity of debris may be large. Different damage mechanisms are expected to operate in these regions, and the most significant mechanisms may change with the degree of wear of the structure.

Due to the complex mechanisms and different phenomena involved in ice abrasion, estimating the deterioration rate of concrete by ice abrasion is particularly challenging. The damage extent and the ice abrasion rate are also related with the travelling/drifting velocity of ice in a particular body of water. This velocity can differ from region to region and from season to season. The rate of ice abrasion has also been reported to increase with ice contact pressure and decrease with ice temperature [112]. Also, the complex nature of ice abrasion of concrete compromises the ability of ordinary abrasion tests to represent uniquely the ice abrasion rate. However, some adapted testing methods have been reported [112]. So far, the ice abrasion damage of concrete is normally managed through provision of low w/c mixtures with SCMs and abrasion-resistant aggregates.

### 2.1.5 Freeze–Thaw Action

Freeze–thaw action on concrete was claimed to be the second main cause of failure of marine concrete in cold regions [97]. The damage mechanisms for concrete due to freeze–thaw action involve both phase change in pores, flow of pore fluids and the associated pressure accumulation. When the pore pressure in concrete during frost actions exceeds the tensile/fracture strength of the solid matrix, cracking will occur. The magnitude of accumulated pore pressure depends on the local temperature, degree of saturation, and material properties such as the porosity, the pore size distribution, entrained air content and tensile strength [41, 52, 66]. On the specimen level, the freeze–thaw damage can be manifested through the developed plastic tensile strain, and this plastic tensile strain accumulates with freezing–thawing cycles [53, 57]. Accordingly, the freeze–thaw damage, if severe enough, can cause degradation in material properties such as strength, stiffness and permeability [135, 151].

During freezing, the pore solution gradually freezes depending on the freezing rate and the pore size distribution, pore connectivity, and pore solution chemistry

[73]. The frozen concrete thus contains a mixture of non-frozen brine and ice [69]. Volume changes during freezing and thawing causes mass flow in pores. Wet frost exposure causes increased water uptake, either by hydraulic pumping or by cryo-suction [52]. To increase the resistance to freeze–thaw damage, marine concrete in cold regions is air-entrained. However, air voids may slowly be filled with secondary precipitated ettringite or water, gradually reducing their mitigating function [41].

The freeze–thaw actions in marine environments involve the seawater. Normally the exposure to both freeze–thaw actions and salt solution increases the risk of surface scaling for concrete, but the presence of seawater also helps to depress the freezing point of pore solution. It has been reported that concrete elements in contact with seawater generated more severe damage compared to samples in contact with pure water after freeze–thaw cycles [2]. Using micro-CT, Maus et al. [93] correlated frost-salt scaling development to the brine volume fraction, which depends on the ice temperature, the solute concentration and the microstructure of the saline ice. Through percolation-based modelling of macroscopic properties, Maus et al. [93] discussed two proposed damage mechanisms, cryogenic suction and glue-spall, for their solute concentration bounds, which are related to the ice growth rate. Besides, the freeze–thaw damage of concrete elements in marine environments also facilitate the subsequent chloride ingress.

### *2.1.6 Electrochemical Process of Corrosion*

The corrosion of reinforced concrete is electrochemical in nature and requires an anode, a cathode, an electrolyte (the pore solution in concrete), and an electrical conductor (the reinforcing bar). The high alkalinity of sound concrete initially offers protection to the reinforcing steel bars. Corrosion of steel in concrete is mainly caused by chlorides and/or carbonation in combination with a high level of environmental moisture. In marine structures, the chloride-induced corrosion predominates [143]. The importance of moisture for active corrosion is discussed in [14, 56, 145] for chloirde-induced corrosion and [15, 50, 55] for carbonation-induced corrosion, respectively.

The corrosion products of steel within concrete vary in volume and solubility. In oxygen-rich environment the corrosion products are typically solid and highly expansive, whereas in marine oxygen-deprived concrete typically soluble and/or low viscosity and low volume products are formed [9, 46]. The loss of steel cross-section in reinforced concrete leads to adhesion loss between steel bars and concrete, reduces the load-bearing capacity of the element, and deteriorates the corresponding structural performance. Additionally, the high-volume corrosion products can generate significant internal stresses in concrete, leading to cracking and damage.

Carbonation-induced corrosion arises because of the neutralisation of the concrete cover, caused by the reaction of carbon dioxide resulting in weak carbonic acids. Once the carbonation front progresses and reaches the steel surface, it can lead to dissolution of the passive layer, potentially initiating carbonation-induced corrosion.

This corrosion, with limited anode-to-cathode ratio, is particularly characterized by general and homogeneous corrosion on steel surface. Chloride-induced corrosion is caused by penetrating chlorides leading to localized break-down of the passive layer and formation of stable localized corrosion zones acting as anodes with respect to the surrounding still passive reinforcement. This results in so-called macro cells with high anode-to-cathode ratio. Unlike carbonation-induced corrosion, the chloride-induced corrosion can lead to substantial and highly localized loss of steel, which can pose significant risks to structural integrity and safety [12, 136].

Oxygen reduction is the typical cathodic process. The pore saturation in concrete may significantly influence the diffusion of oxygen within the concrete matrix. High saturation levels reduce the oxygen content and diffusion, thereby limiting the progress of steel corrosion [108] but not necessarily hindering it [10]. On the other hand, high saturation levels reduce also the electrical resistivity of concrete thus promoting the corrosion [4, 71]. The impact of the partial processes, anodic, cathodic and ohmic, on the carbonation-induced corrosion and the chloride-induced corrosion was discussed comprehensively in literature [3, 4, 61, 124].

Extended exposure to seawater progressively leaches the cementitious materials within the concrete matrix [44, 67]. This leaching process increases the porosity and permeability of the affected zones, and elevates the risk of corrosion for the reinforcing steel embedded in low-quality concrete and/or with low cover thickness [38]. The combined mechanism of chloride ingress and carbonation on corrosion is described in [3, 92]. The combined phenomenon has not yet been thoroughly understood, and Alonso and Andrade [3] found a corrosion rate of steel depending on both humidity and cement type in mortars contaminated both by carbonation and chlorides.

### *2.1.7 Biological Processes*

Concrete in coastal areas can be exposed to biological processes that can lead to physical, aesthetic and/or chemical deterioration. Biological degradation is caused by marine biota that grows on concrete surfaces [117]. Mechanisms such as carbonation, and extensive wet/dry cycles in intertidal areas, lower the pH of concrete and increase the relative humidity, thus creating a favorable environment for microorganism metabolic activity [149]. Organisms such as bacteria, fungi, algae and lichens can erode concrete surfaces, increase porosity, and promote crack formation, thereby potentially reducing the structural integrity and service life of a structure. The metabolic activity of such organisms can lead to the solubilization of metal ions and the decomposition of the cementitious substrate. Fungi and bacteria, such as nitrifying bacterial and sulphur-oxidising bacteria, can also produce inorganic and organic acids that lead to corrosion of the reinforcement, and also react with the concrete to form gypsum.

**Fig. 2.4** Intertidal concrete structures, Pembroke Port, Wales, United Kingdom. Courtesy of Dr. Riccardo Maddalena

It has been reported that oxygen is one of the key-parameters for the growth of microorganisms. In concrete permanently submerged, microbial activity is proportional to the depth of the structure underwater; in less deep waters, dissolved oxygen might be enough to allow microorganism growth and lead to formation of aggressive substances. This can be the case with oil storage tanks, tunnels and foundations [47]. In the SPL and TID zones, the deterioration of concrete is mainly due to seawater salts, and development of biofilms on the concrete surface can limit further deterioration by reducing chemical corrosion that would normally happen in the SPL zone, cf. Fig. 2.4. It should be noted growth algae and related microorganisms on surfaces of concrete structures can impede visual inspection.

However, growth of algae can contribute to the depletion of calcium hydroxide in concrete for metabolic activity, leading to the production of metabolic acids and subsequent degradation of the pore structure [18, 68]. Other than surface damage, microorganism can also penetrate the cementitious matrix through microcracks and capillary action, generating tensile stresses that increase the crack size and worsen the concrete porosity. In turn, microorganisms can accelerate corrosion of rebar reinforcement by their production of organic and inorganic acids. A visual summary of reinforced concrete deterioration is provided by Sanchez-Silva and Rosowsky [127] in Fig. 2.5.

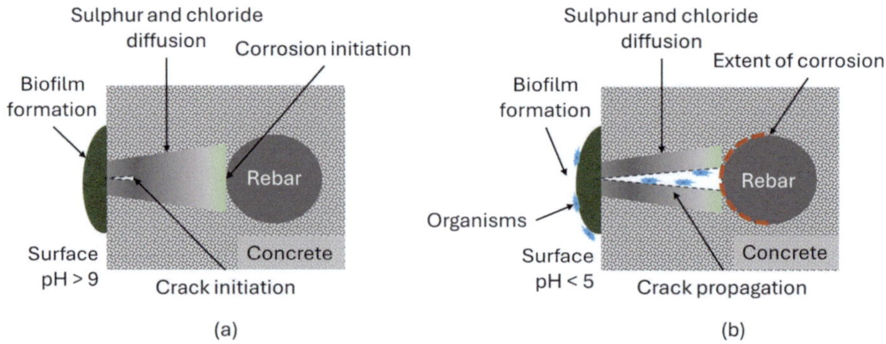


**Fig. 2.5** Mechanism of bio-deterioration of concrete; **a** surface pH reduction, chloride and sulphur diffusion, crack initiation and corrosion initiation, **b** crack propagation, accumulation or micro-organisms, corrosion propagation and delamination. Illustration adapted from [127]

## 2.2 Environmental Factors for Marine Exposure

### *2.2.1 Mechanisms*

The environmental factors describe the environmental and meteorological envelope of structural concrete. These factors, for marine environment, are closely related to the hydrology of seawater and can include its salinity, the temperature, and the drying-wetting cycles or freeze–thaw cycles acting on the concrete surface. The salinity of seawater is defined as the total content of soluble ions such as $Na^+$, $Cl^-$, $Mg^{2+}$, $SO_4^{2-}$. These aqueous ions participate in the diffusion, capillary absorption, and chemical reactions with cement hydrates. The concentration of these ions in seawater evolves with the hydrological and meteorological actions such as precipitation and evaporation, inflow of fresh water and melting of glaciers [102]. The seawater temperature depends on the local climate and impacts on the ionic activities, changing the ionic diffusion rate and the chloride adsorption [95]. The drying-wetting cycles may result from the periodical contact of concrete surfaces with air and seawater, or the intermittent atmospheric precipitation. Under the cyclic action external ions and agents transport into concrete mainly from the capillary absorption (suction), and its intensity depends on the pore saturation of concrete and the respective periods of drying and wetting [79]. As the concrete surface is exposed simultaneously to seawater wetting and frost climates the freeze–thaw cycles become active in marine environment. Under such conditions, apart from the possible freezing damage the "pump-effect" can absorb the external seawater directly into concrete [132] and increase the sorptivity [48] with the number of cycles as the crucial factor for chloride penetration [83]. In parallel, the concrete surface can also be subject to erosion, by flowing seawater, and abrasion by flowing sediments and/or ice in frost climates.

Apart from seawater, the presence of moisture and $CO_2$ in the atmosphere is also relevant to the durability of structural concretes in marine environments. The ingress of corrosive agents is highly sensitive to the moisture level or (degree of) pore saturation in concrete, leading to variable degradation extents in marine exposure zones. The combined effect of chlorides and moisture is particularly pronounced in the ATM zone. Also, the atmospheric $CO_2$ always jointly, with marine chlorides, contributes to the durability processes for urban infrastructure.

### 2.2.2 *Observations from Literature*

The temperature and salinity of seawater will affect the ionic diffusion into concrete and the chloride penetration can be promoted by a higher temperature or a higher salinity. Lindvall [87] performed a systematic study on the effect of seawater temperature and salinity on the chloride ingress, in terms of the apparent surface chloride concentration, $C_{sa}$, and the apparent chloride diffusion coefficient $D_{F2}$. Concrete specimens were exposed at different exposure sites varying in annual mean salinity (AMS) of seawater: BS + OS (exposure in the Baltic Sea and in Øresund, AMS 3–5 g/l), KG (exposure in Kattegat, AMS = 10 g/l), KV (exposure in Kjøpsvik, AMS = 17 g/l), Rest (other exposure locations, AMS = 18–23 g/l) and Lab (exposure in the laboratory. AMS = 5 g/l and 20 g/l). The results showed that the temperature of seawater water had significant effect both on $C_{sa}$ and $D_{F2}$, an increasing temperature decreased $C_{sa}$ but increased $D_{F2}$, both by more than 100% with temperature rising from 5 to 25 °C. The salinity of the seawater of different locations had less effect on $C_{sa}$ and $D_{F2}$.

The different contact conditions between seawater and concrete surface (see description of exposure zones in Sect. 1.3) provide different values for the apparent chloride diffusion coefficients typically following the order SPL < TID < SUB. The results from Zhanjiang exposure station [42], showed that for concrete specimens with w/c = 0.55 and exposure age of 10 years, the regressed apparent chloride diffusion coefficients were 6.5 $10^{-12}$ $m^2/s$ (SPL), 23.3 $10^{-12} m^2/s$ (TID) and 26.0 $10^{-12} m^2/s$ (SUB).

In marine environment, the distance from coastline will impact on the chloride deposit rate and quantity on the concrete surface. Meira et al. [98] investigated the correlation between the chloride deposition on concrete surface from marine aerosol and the chloride content in concrete (surface layer) at distances from coastline from 10 to 200 m. The results showed that the chloride content in the concrete was related to the chloride deposition and the surface chloride deposit quantity decreased rapidly from the coastline, the chloride deposit rate changed from 453 mg/($m^2$day) at 10 m to 104 mg/($m^2$day) at 100 m and only 16.4 mg/($m^2$day) at 200 m.

The reach of the marine aerosol containing chlorides is determined by the aerodynamics created by the local climates. Among these aerodynamic parameters, the wind speed has been found to be the key factor. Morcillo et al. [111] investigated the correlation between the wind speed and the chloride deposit rate on the Spanish coast of Tarragona. The results showed that the numbers of hours of wind with speed above 3.0 m/s is closely correlated to the deposit rate, and 27 h of wind with speed above 3.0 m/s brought the deposit rate of chlorides as high as 110 mg/($m^2$day) in the test location.

For marine environments with frost climates, the environmental parameters involve, among others, the freezing temperature, the cooling rate, the number of freeze–thaw cycles and the moisture condition [54]. The additional water supply from the ambient environment plays an important role for the damage accumulation [40, 85].

## 2.3 Material Factors for Marine Exposure

### *2.3.1 Mechanism*

The material factors of concretes in marine exposure refer to the composition of raw materials, the proportioning of mixtures, and the properties of concrete in fresh and hardened states. The concrete binder includes Portland cement, supplementary cementitious materials and other fillers. The mineral composition of binder materials determines the hydration products and pore solution chemistry, providing the internal environment for ionic transport and reactions. Among the proportioning parameters, the water-to-binder (w/b) ratio is central since it can determine the possible hydration extent of the binder and shape the internal pore structure. The formed pore structure, e.g., pore volume and tortuosity, will in turn determine the main transport properties of concrete such as the gas permeability, ion (chloride) diffusivity, water permeability and water absorption rate [1]. The near-surface zone of concrete should be handled separately because this zone is different from the bulk concrete. This zone is rather sensitive to the concreting and curing practice [60] forming the weakest link against the chloride penetration [11, 23].

### *2.3.2 Observations from Literature*

#### 2.3.2.1 Binders

Bérubé and Carles-Gibergues [21] showed that the cement clinker substitution by pozzolan in cementitious materials led to the formation of a gel resulting from the reaction of pozzolan with cement hydration products (pozzolanic reaction), which contributes to reduce the material's permeability and its chloride diffusion coefficient. Additional studies [35, 115, 140] conducted on pastes containing blast furnace slag or fly ash revealed a lower diffusion coefficient compared to pastes without mineral additions. This phenomenon may stem from the alteration of the pore structure of the pastes due to the formation of C-S-H, CAH and CASH, resulting in a much finer microstructure. Dhir et al. [36] observed that the substitution of cement by 33.3%,

50%, or 66.7% blast furnace slag (at w/c = 0.55) influenced the chloride fixation capacity in cement pastes, which increased with the increase in substitution rate. In a comparison between slag and fly ash effects on chloride fixation, [13] showed that replacing 15% by mass of fly ash with slag increased the chloride binding capacity in a cement paste with w/c = 0.5. However, for a w/c ratio of 0.7, replacing fly ash with slag decreased the content of bound chlorides. Incorporating silica fume in cementitious materials Bérubé and Carles-Gibergues [21] observed a decreased chloride binding capacity.

Several researchers [35, 120, 150] highlighted that the chemical composition of the pore solution of cementitious materials impacted the chloride diffusion and the chloride binding in the cement matrix. Delagrave et al. [35] demonstrated that high concentrations of $OH^-$ and $SO_4^{2-}$ in the pore solution reduced the chloride binding capacity of the material and the $OH^-$ and $SO_4^{2-}$ reacted more readily than chlorides with the still anhydrous cement to form sulfo-aluminates. This explains the heightened resistance of cements rich in $C_3A$ in chlorinated environments only. Conversely, their resistance in seawater was lower due to their sulfates content. Figure 2.6 illustrates the influence of sulfates content in the pore solution on chloride binding and shows that the chloride binding capacity in an OPC cement paste increased with decreasing sulfates content. Xu [150] also found that the chloride binding capacity depended on the composition of the solution in contact with the material. A higher chloride content was bound when they originated from $CaCl_2$ rather than NaCl. This elucidates the dependence of the isotherms on the associated cation, as chlorides were bound more with calcium than with sodium. This higher proportion of free chlorides when the cation of chloride is sodium than calcium was also found by other authors [8, 147, 109, 110]. Additionally, $CaCl_2$ was found to reduce the portlandite solubility and the pH of the pore solution, resulting in less competition between $OH^-$ and $Cl^-$ and an increase in chloride binding capacity of the material [8, 137].

Sørensen et al. [139] compared the chloride ingress of 15 different concrete mixtures with nearly the same water-to-binder ratio of 0.40. From the regression on chloride profiles from exposures of 0.5 year and 2 years, the magnitude of regressed apparent chloride diffusion coefficient observed the order: slag cement > fly ash (FA) cement > silica fume (SF) cement > ordinary Portland cement (OPC). Investigating six of the concretes after 10 years exposure, [44] confirmed the trend in chloride ingress resistance and observed a correlation between the location of maximum chloride content and the thickness of the microstructurally changed zone as well as a variation in the depth of leaching, which has been found to affect the chloride ingress depth [91]. In another marine exposure test, Thomas and Matthews [142] used their 10-year results to show the correlation between the steel loss due to corrosion, the fly-ash content in binder (0–50%) and the concrete strength grade (C25-C45). From the steel loss data, the corrosion extent, for each concrete strength grade, decreased with the fly-ash content. Other results from different exposure practices confirmed the increase of concrete resistance against the chloride ingress by incorporating 5–10%SF [128], 15–35%FA [30] and various other SCMs [121]. Mohammed et al. [104] studied the performance of the alumina-cement concrete and found, compared to Portland cement concrete, the surface had a denser microstructure (smaller porosity),

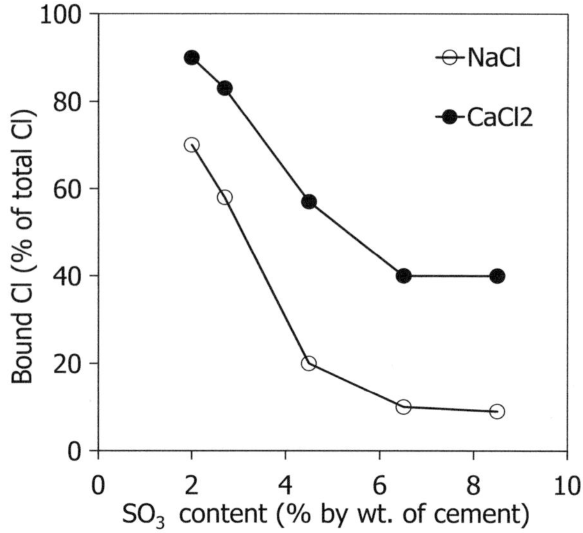


**Fig. 2.6** Influence of $SO_4^{2-}$ content in the cementitious material on chloride binding (salts in two forms NaCl and $CaCl_2$) in ordinary Portland cement paste. Illustration adapted after [150]

higher resistance to the chloride ingress (elemental distribution) but a coarser pore structure on the steel-matrix interface.

#### 2.3.2.2 Water-To-Cement Ratio

From the viewpoint of concrete mixture proportioning, the water-to-cement (w/c) ratio or water-to-binder (w/b) ratio has been always the key factor. Multiple studies have confirmed its dominating and limiting effects on the resistance of concrete to chloride ingress. Chalee and Jaturapitakkul [26] conducted exposure tests in marine environment (TID zone in Chonburi, Thailand) on concrete mixtures with different binder compositions and w/b ratios from 0.45 to 0.60. The results after 5-year exposure tests were reported: mixtures with lower w/b ratios had lower values of apparent chloride diffusivity $D_c$, particularly for the mixtures without fly ash. Decreasing w/b ratio from 0.65 to 0.45, the values of $D_c$ of the cement concretes at 5-year exposure were decreased from $5.6 \times 10^{-12}$ to $3.1 \times 10^{-12}$ $m^2/s$ while the $D_c$ of the concretes containing 35% original fly ash decreased from $1.2 \times 10^{-12}$ to $0.8 \times 10^{-12}$ $m^2/s$. Further, the w/b ratio was more effective on reducing the chloride diffusion coefficient for cement concretes than for concretes incorporating fly ash in binder. Concerning the resistance to ice abrasion, Laplante et al. [77] and Dhir et al. [37] showed the positive impact of a decreasing w/c ratio. This was explained by the lower porosity, better bond and increased compressive strength of low w/c concretes. For example, a 45% higher abrasion loss was observed for on a concrete with the w/c ratio of 0.36 compared to a concrete with w/c ratio 0.28 [77].

#### 2.3.2.3 Aggregates

The aggregates are essential components for concrete materials. Thomas and Bremner [141] investigated the impact of light-weight aggregates, from expanded shale, in CANMET concrete blocks on the chloride ingress. The concrete blocks were exposed on the Treat-island site and retrieved for analysis after 25 years' marine exposure. The w/b ratios were controlled as 0.40–0.60 and the slag content in binder was from 0 to 65%. For comparison, concrete specimens with normal aggregates were also tested as reference. The general observation on the chloride profiles was that the LWA-concrete had notably higher chloride content without slag in binder and the chloride profiles of LWA-concrete were very near to the normal aggregate concretes for binders incorporating slag. This may be explained by the internal curing effect of LWA, which improved the microstructure or pore structure of concrete during the hardening process. Concerning the resistance to ice abrasion, the properties of aggregates, such as hardness and maximum size, and aggregates amount have been reported as important factors [131]: the weaker the aggregates, the greater the abrasion loss. Dhir et al. [37] investigated the effect of maximum aggregate size, ranging between 5, 10, 20 and 40 mm with constant compression strength and aggregate volume, on the abrasion resistance of the concrete and found a decreasing abrasion resistance with increasing aggregate size. In contrast, [89] found, for low-strength concretes with $w/c = 0.50$, the abrasion resistance increased with the maximum aggregate size. This was explained by the larger aggregates causing a reduced percentage of surface area fraction of the (weaker) cement paste and that larger aggregates required more energy to dislodge. In high strength concrete with $w/c = 0.28$, no such correlation was apparent with aggregate sizes [89].

#### 2.3.2.4 Specimens (Protection)

As the concrete specimens are placed into the exposure sites, the physical condition also plays important role in the exposure practice. Schueremans et al. [129] reported their results on the chloride ingress from extracted cores from concrete surface protected by silane impregnation (alkyltriethoxysilane) in tidal zones. The general observation was that the hydrophobicity of concrete surface was still valid after 12 years' exposure to tidal actions, the chloride ingress had been limited by the surface treatment of silane impregnation, and the treated surface presented a peak on the chloride profiles within the depth of 20 mm from the surface. Huang et al. [62] reported the chloride ingress into concrete specimens bearing polyurethane coatings in marine exposure conditions after 19 years' exposure in splashing zones in Zhanjiang exposure station. The chloride ingress into concrete specimens with different coatings was, at all depths and exposure ages, much lower than the control specimens without coatings. For the depth 1–3 cm, the accumulated chlorides were 7.3–24.9 times, 18.1–45.3 times lower in coated specimens compared to control specimens for exposure ages of 5 years and 8 years respectively.

## 2.4 Loading and Service Factors for Marine Exposure

Structural concretes work under service loadings. The loading, if excessive, can induce cracking in concrete and facilitate the ingress of external aggressive agents. The impact of loading on durability processes is principally due to the induced cracking [101]. Cracks with moderate opening, e.g. within 0.3 mm, were found to have rather limited influence on the corrosion risk of embedded steel in concrete [130, 158]. However, under different circumstances, crack openings down to 0.1 mm were found to increase the corrosion risk: Castro-Borges et al. [25, 24] found a clear correlation between crack openings in the range 0.1–0.8 mm and the pit depth into steel bars. Cracking due to severe early-age or drying shrinkages was judged more harmful for steel corrosion [96]. Still, more field and laboratory observations are needed to confirm these statements. Under accidental actions, e.g. ship impact or release of aggressive chemicals for port structures, the induced damage can accelerate the corrosion of embedded rebars [94].

### *2.4.1 Mechanical Effects on Transport*

Under service loadings, the principle of reinforced concrete includes the service cracking of concrete and the steel still working within the elastic limit. Accordingly, the effects of mechanical loading on the deterioration kinetics of concretes can involve both the pure loading effect, without material damage, and the intervention of cracks, in damage phase of concrete. The consequence of pure mechanical loading on concrete is the elastic deformation of its solid matrix, which can be quantified through general poromechanics. Li [78] demonstrated that un-realistic high stress, at least 85 MPa, is needed for a notable change of concrete porosity. Since the service stress in structural concretes is far lower than this value, the pure influence of stress on the alteration of transport properties can be neglected. In other terms, the change of the transport properties under loadings must be attributed to the cracking.

The alteration of transport properties of concrete materials by various loadings has been extensively investigated in the literature [17, 22, 86, 123, 152]. The alteration of transport properties was measured from the specimens either under loading or after loading. Although these data were obtained from different experimental procedures, a common observation was that the loading level had little impact on the transport properties as compressive stress remaining inferior to 0.7–0.8$f_c$ while for tensile loading cases this threshold value was observed to be as low as 0.3$f_c$. Normally, the allowable compressive stress in concrete specified in design codes can rarely be higher than 0.5$f_c$. So, a basic judgement is that the service compressive stress in structural concretes will not be high enough to cause notable change to transport properties. Nevertheless, the long-term effect of stress on the transport properties, e.g. via creep, remains to be quantified further.

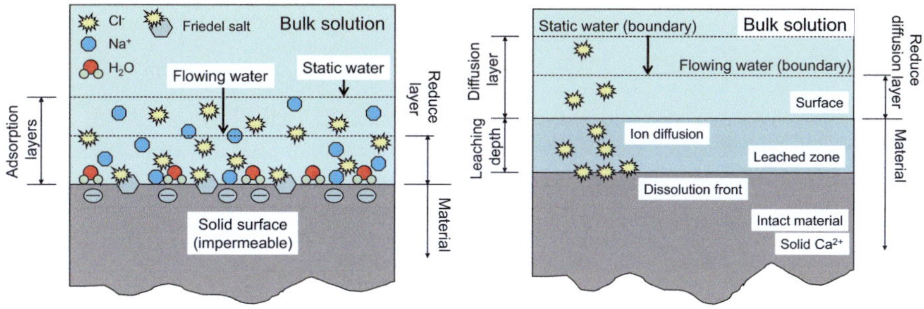


**Fig. 2.7** Impact of flow on the mass exchange in crack: flow-reduced surface adsorption (left) and flow-enhanced surface leaching (right). *Source* Li and Li [81], Fig. 4. Reproduced with permission of Elsevier

## 2.4.2 Single Crack

A single crack can facilitate the liquid penetration into cement-based materials [81]. Two observations were significant [84]: first, the hydraulic equivalent width is far less than the real geometry crack width; second, the liquid flow takes a transient behavior, and the flow rate is much larger in the first few hours and then decreases substantially with time. The surface roughness is responsible for the difference between the equivalent and real openings. The transient flow in the cracks was attributed to the secondary hydration of cement [39]. However, the repetitive flow tests through a same crack [84] revealed that this transient nature could be purely physical due to the friction between the flowing liquid and the rough surfaces.

When liquids flow across the fracture surfaces of a single crack in concrete, mass exchange (dissolution and precipitation) and transfer (transport) can occur. Li and Li [81] investigated the adsorption of chlorides in flowing water by the fracture surface and the leaching of soluble matters on the fracture surfaces by the flowing water. The flow rate of water solution containing chlorides was found to decrease both the chloride adsorption rate and capacity of fracture surface. The reason was attributed to the intervention of liquid flow in the multi-layer adsorption of chlorides on the fracture surface cf. Fig. 2.7 (left). From the same study the leaching rate of the soluble matters, $Ca(OH)_2$, was observed to augment notably by flowing water compared to the static leaching case. The underlying mechanism is the removal of diffusion layer thickness, by the liquid flow, on the fracture surface for the dissolved ions, cf. Fig. 2.7 (right).

## 2.4.3 Crack Network

For mass transport in concrete containing multi-cracks, two distinct features are introduced: the interference among the cracks, and the local connectivity of cracks, cf.

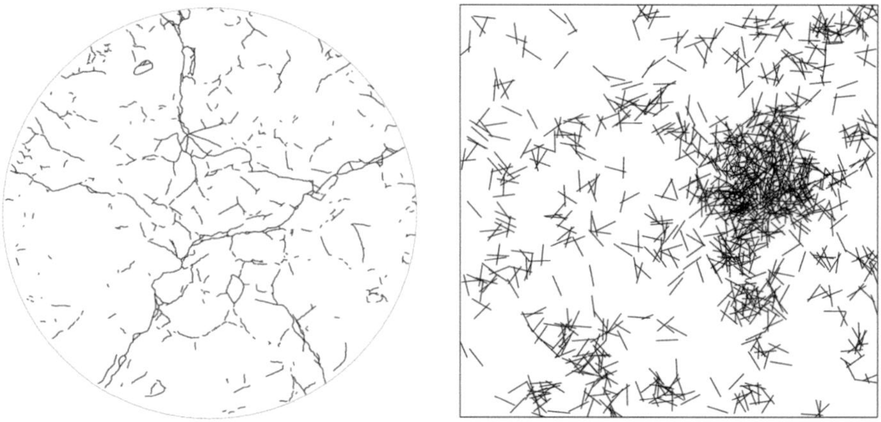

**Fig. 2.8** Distribution of crack networks: (left) cracks created by axially compressive loading in HVFC5 specimen [154], and (right) crack network generated numerically with connectivity f = 0.367 [80]. Reproduced with permissions of ASCE (left) and Springer (right)

Fig. 2.8. These two features can be handled either from average quantities of multi-cracks [155, 157] or from the percolation of crack clusters [80]. The average quantities include the crack density, crack orientation factor and the global connectivity while the percolation description of crack clustering proposed a main-cluster-based connectivity considering the main cluster size, the probability of cracking belonging to the main cluster and the fractal properties of the main crack cluster. Based on these descriptions, the alteration of concrete transport properties by multi-cracks are considered by two different theories [81].

The effective medium theory (EMT) regards cracks as inclusions in the homogeneous matrix and applies mathematical homogenization to express the altered transport properties in terms of the crack characteristics [133]. So far, the EMT has had multiples major successful applications for engineering materials [58]. However, the assumption of each crack contributing equally to the transport does not reflect the physical reality, leaving out the crucial effects of the connectivity and interaction among the cracks. As the theoretical counterpart, the percolation theory addresses directly the crack clustering during the mass transfer, especially as the spanning cluster is formed. The EMT is more suitable for describing the multi-cracks evenly distributed and weakly connected, and the percolation theory fits for highly clustered cracks near percolation (total connection of cracks). The mathematical equations for the altered transport properties and the relevant scaling law can be found in the literature [81, 82, 156].

## 2.5 Performance and Consequence Factors for Marine Exposure

Under environmental actions, the relevant durability processes reviewed in Sect. 2.1 can cause two fundamental changes to the structural concretes. For near-surface concrete, both the solid matrix and the pore solution may be notably changed due to the mass exchange between the environment and the material. The affected properties can include mechanical properties abrasion resistance, frost resistance and transport properties. For the bulk concrete, the mass transfer across the cover will supply aggressive ions, such as chlorides and sulfates, inducing chemical and electrochemical (corrosion) reactions. And, the corrosion of the embedded steel will affect the performance and the structural bearing capacity [12, 143].

### *2.5.1 Microstructure and Chemical Composition*

Mohammed et al. [104] investigated comprehensively the microstructure, mineral and chemical composition of concrete specimens in marine sites after 30 years of exposure. The microstructure was observed by SEM (Scanning Electron Microscopy) for the surface zone, the inner zone, the matrix-aggregate interface and the matrix-steel interface. The pore structure was characterized through the porosity and pore size distribution. It was found that, after 30 years of exposure, the surface zone of SCB (Type B slag cement) and AL (alumina cement) specimens presented a dense microstructure, and the matrix-aggregate interface was also denser. But the microstructure was less dense for AL specimens due to the mineral phase transformation (from $CAH_{10}$ to $C_3AH_6$) with the same observation for matrix-steel interface. The mineral composition was characterized using XRD (X-ray diffraction) and the elemental distribution was detected through EPMA (Electron probe micro analyzer). The sulfates and other ions except chlorides were rather limited in the specimens and no Friedel's salt was found in AL specimens and the chlorides were assumed to be absorbed into the interlayer of the hydrates.

Kwon et al. [75] exposed concrete specimens in the exposure station, located in Tuticorin new harbor area on the southeast coast of India with water depth of 5 m. Three binders were chosen, OPC (ordinary Portland cement), PPC (Portland Pozzolana cement) and PSC (Portland slag cement), for concrete mixtures with w/b = 0.55 and 28 day strength around 30 MPa. These concrete specimens were placed into SUB, SPL and ATM zones for 10 years. For the chemical composition, the pH value was measured on the extracts of concrete specimens and XRD was used for the mineral analysis. The pH value measurement showed that important leaching occurred for concrete specimens in SUB and SPL zones, and the XRD analysis showed that Friedel's salt formed more in SUB and SPL zones than in ATM zone due to the higher quantity of chlorides penetrated.

### 2.5.2 Mechanical Properties

The compressive strength of concrete materials is the most valued property by engineers due to its fundamental nature and its highly standardized test method. Thus, the compressive strength is also retained as the first indicator or factor for the impact of environmental actions on the concrete properties. Chalee et al. [27] investigated the change of compressive strength of concrete specimens under marine exposure for 5 years. The binder incorporated GRBA (Ground rice husk-bark ash) with contents of 0–50 wt.%. After 5 years' exposure, the compressive strength was compared to the strength measured at 28d in laboratory. Generally, the compressive strength did not change notably after exposure, concretes with 0% GRBA showed a decrease of 5% for compressive strength, and all concrete with GRBA gained 10% in compressive strength. In some early field investigations, the compressive strength was investigated for concretes using different binders of ordinary Portland cement (OPC), high early-strength cement (HES), Type B slag cement (SCB) and alumina cement (AL) with w/c ratio between 0.52 and 0.53 [104]. After 20 years' exposure, the concretes of OPC, HES and AL only showed a slight decrease [45] while other concretes with fly ash cement, slag cement and OPC showed no reduction of strength after 15 years' exposure except the concrete using fly-ash cement [107]. However, the concrete cores extracted from exposed beam elements after 23 years' exposure showed significant strength increase [106].

### 2.5.3 Transport Properties and Crack Healing

The transport properties of concretes can be changed through material deterioration and/or cracking during the exposure in marine environments. The cracks formed in working state of reinforced concrete elements can have impact on the transport properties and the durability performance in marine environment. Figure 2.9 reports experimental results on transport and durability properties of concrete altered by cracking. The cracking was made on cubic specimens of concrete either by axial compressive loading or by drying-wetting cycles. The cracking extent of specimens was quantified by ultrasonic pulse velocity (UPV) and the crack patterns were observed by scanning electron microscope. The cracked specimens were subject to rapid chloride permeability test and rapid freeze–thaw test. The correlation between cracking extent and the altered properties was established through a unified damage factor. Again, the role of the cracking pattern was highlighted from the results, cf. Fig. 2.9.

The cracks formed in reinforced concrete can be healed with time when exposed to marine environment. Investigating self-healing of cracks after 25 years of exposure in three marine concretes differing in binder composition (4 and 12 wt.% SF, 0 and 20 wt.% FA), Danner et al. [34] observed the general feature of mineralogical sequence of self-healing products, see Fig. 2.10. Calcite was observed in the outer 0–5 mm of

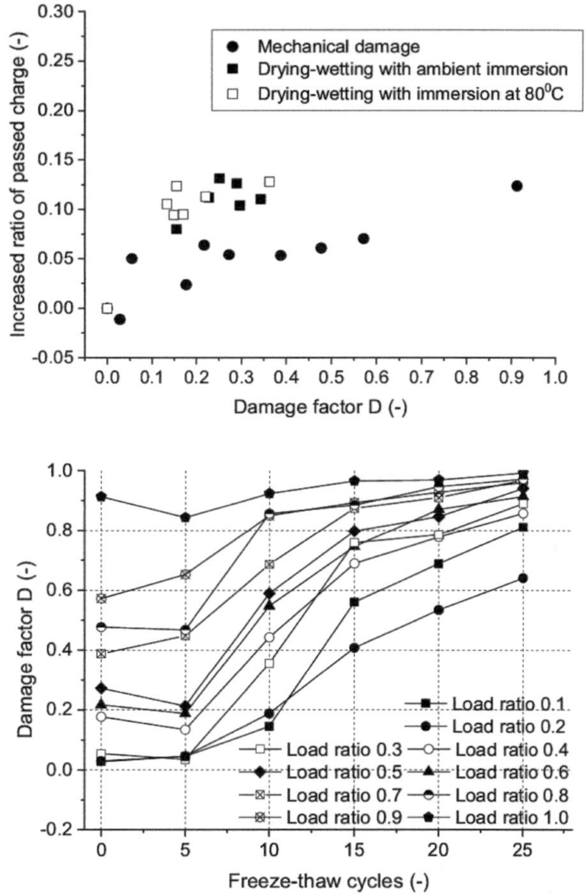


**Fig. 2.9** Alteration of chloride migration (measured by ASTM C1202 method) in terms of the damage factor quantified by UPV for different sources of cracking (top) and the evolution of damage factor with the freeze–thaw cycles according to ASTM C666 test (bottom)

the crack followed by brucite, partly intermixed with calcite, to a depth of 30 mm. At greater depth only precipitation of ettringite was observed. It was hypothesized that the self-healing of cracks exposed to seawater takes place in the outer part of crack by precipitation of ions from the seawater, partly in reaction with ions from the cement paste, and at larger crack depths by dissolution and reprecipitation of ettringite. Cracks of width up to 0.2 mm appeared closed, while wider cracks were partly filled after 25 years of submerged marine exposure in Northern Norway [34].

## 2.5.4 *Electrochemical State of Embedded Steel*

The corrosion behavior of embedded steel in concrete is one of the ultimate performances sought by the engineering community. Kwon et al. [75] exposed concrete

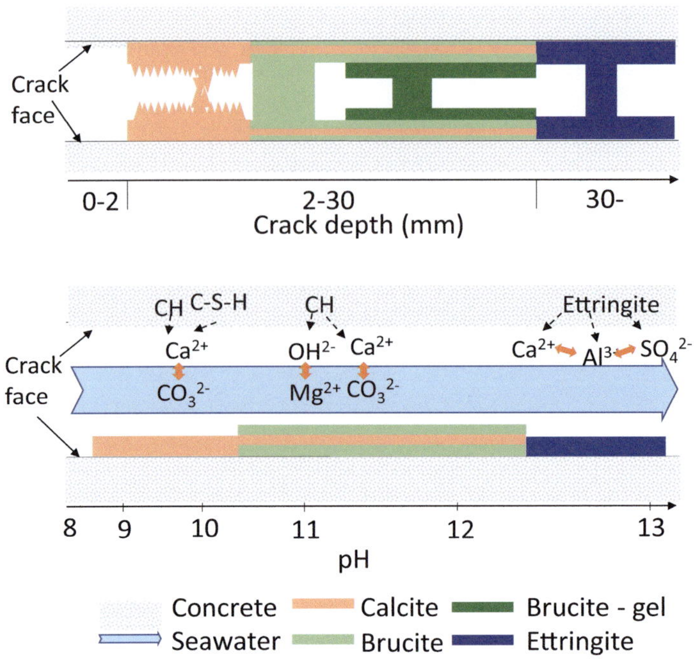


**Fig. 2.10** Features of mineralogical sequence of self-healing products: (top) observed mineral sequence of self-healing products and (bottom) proposed self-healing mechanism. Image was originally published in [34] and is licensed under a Creative-Commons license CCBY 4.0

specimens embedding TMT (Themo Mechanically Treated) steel bars, the specimens were placed into ATM, SPL and SUB zones for 10 years, and the electrochemical states were measured using AC impedance and potentiodynamic polarization methods. Results from both measurements showed the TMT steel had higher corrosion current intensity in OPC concrete than concretes with SCM (pozzolana or slag). For a same specimen, the corrosion current was larger in the SPL zone than the ATM and SUB zones. Andrade [7] exposed bars embedded in different concrete mixes made with ordinary Portland cement during more than 20 years at a beach of the Atlantic Ocean, finding that the corrosion was similar irrespective of the amount of cement.

In addition to the global exposure and concrete composition, the electrochemical state of embedded steel can be influenced by the cracks and other defects/inhomogeneities. Mohammed et al. [105] studied the healing of cracks in specimen in marine exposure sites and its impact on the corrosion behaviors of embedded steel bars after 15 years' exposure in a tidal pool. The micro-cell current density was measured for specimens bearing pre-fabricated cracks of widths 0.1–0.5 mm. For crack widths within 0.4 mm, both self-healing and reduced corrosion rate were observed. However, the possible link between initial cracks and corrosion in de-iced elements still remains unclear, as outlined in a recent review [70] which synthesized the peculiarities of relevant investigations based on the state-of-the-art of understanding.

## 2.6 Chapter Summary

The mechanisms affecting the durability of concrete materials and structures in marine environment were reviewed. The physical processes, due to the direct contact with seawater, include freeze–thaw action and ice abrasion of the concrete surface. The chemical processes, due to the direct contact with seawater or/and the atmosphere, include carbonation, chloride ingress and binding, and other reactions with seawater. In addition, the electrochemical process of steel corrosion and biological processes causing surface growth are of engineering concern for concrete structure in marine environments. Many of these processes can be coupled. In ATM zone, the concrete surface is subjected to chloride deposition and carbonation simultaneously. In SPL zone, the concrete surface is exposed to strong moisture transport due to drying-wetting action causing chloride accumulation at the depth of the convection zone. In TID zone, the concrete surface is subject to ingress of chloride and other ions in the seawater, leaching, erosion by flowing water, in frost conditions freeze–thaw action and ice abrasion. The concrete surface in SUB zone has the same processes as TID, except for freeze–thaw action.

The marine environment affects the durability of concrete materials and structures via the environmental, material and service state factors. These factors were reviewed for both the involved mechanisms and the obtained research results, which can be used further to assess the qualitative correlations between these factors (environmental, material and service) and the durability performance. Among the environmental factors, the drying-wetting action is a primary factor which has impact on all the long-term performance of structural concrete and increases the corrosion risk of the embedded steel bars. For the material factors, the concrete compactness has the most impact on the durability performance. For the service factors, not the straightforward loading level but the resulted cracking is responsible for the altered durability performance.

## References

1. AFGC: Durabilité des bétons : Méthode recommandée pour la mesure des grandeurs associées à la durabilité. In: Compte-rendu des Journées Techniques AFPCAFREM, vol. 283 (2007)
2. Achintya, Prasad, M.M.: Behaviour of concrete in freeze-thaw environment of sea water. J. Inst. Eng. India Civ. Eng. Div. **84**, 96–101 (2003)
3. Alonso, C., Andrade, C.: Corrosion of steel reinforcements in carbonated concrete containing chlorides. Adv. Cem. Res. **1**(3), 155–163 (1988)

4. Alonso, C., Andrade, C., González, J.A.: Relation between concrete resistivity and corrosion rate of the reinforcements in carbonated mortar made with several cement types. Cem. Concr. Res. **18**, 687–698 (1988)
5. Amiri, O., Aït-Mokhtar, A., Dumargue, P., Touchard, G.: Electrochemical modelling of chloride migration in cement based materials part I: theoretical basis at microscopic scale. Electrochimica Acta **46**, 1267–1275 (2001)
6. Amiri, O., Friedmann, H., Aït-Mokhtar, A.: Modelling of chloride-binding isotherm by multi-species approach in cement mortars submitted to migration test. Mag. Concr. Res. **58**, 93–99 (2006)
7. Andrade, C.: Steel corrosion rates in concrete in contact to sea water. Cem. Concr. Res. **165**, 107085 (2023). https://doi.org/10.1016/j.cemconres.2022.107085
8. Andrade, C., Page, C.L.: Pore solution chemistry and corrosion in hydrated cement systems containing chlorides salts: a study of cation specific effects. Br. Corros. J. **21**(1), 49–53 (1986)
9. Andrade, C., Tavares, F., Toro, L., Fullea, J.: Observation on the morphology of oxide formation due to reinforcement corrosion. In: Proceeding Joint fib-RILEM Workshop "Modelling of Corroding Concrete Structures". Madrid, Spain, RILEM Book Series, vol. 5, pp. 179–193, November 2010 (2010)
10. Andrade, C., Fullea, J., Toro, L., Martinez, I., Rebolledo, N.: Reinforcement corrosion in chloride media in absence of oxygen. In: NUCPERF 2012, 12–15 November 2012, Cadarache, France (2012). https://doi.org/10.1051/epjconf/20135606002
11. Andrade, C., Díez, J.M., Alonso, C.: Mathematical modeling of a concrete surface "skin effect" on diffusion in chloride contaminated media. Adv. Cem. Based Mater. **6**, 39–44 (1997). https://doi.org/10.1016/S1065-7355(97)00002-3
12. Andrade, C., Rodriguez, J.: A validated User's Manual for assessing the residual life of concrete structures. Project CONTECVET (IN309021), GEOCISA and Torroja Institute, Madrid, Spain (2001). https://www.ietcc.csic.es/wp-content/uploads/1989/02/manual_contecvet_ingles.pdf
13. Angst, U., Elsener, B., Larsen, C.K., Vennesland, Ø.: Critical chloride content in reinforced concrete — a review. Cem. Concr. Res. **39**, 1122–1138 (2009). https://doi.org/10.1016/j.cemconres.2009.08.006
14. Angst, U., Isgor, O., Hansson, C.M., Sagues, A., Geiker, M.: Beyond the chloride threshold concept for predicting corrosion of steel in concrete. Appl. Phys. Rev. **9**, 011321 (2022). https://doi.org/10.1063/5.0076320
15. Angst, U., Moro, F., Geiker, M., Keßler, S., Beushausen, H., Andrade, C., Lahdensivu, J., Köliö, A., Imamoto, K., Von Greve-Dierfeld, S., Serdar, M.: Corrosion of steel in carbonated concrete: Mechanisms, practical experience, and research priorities – a critical review by RILEM TC 281-CCC. RILEM Techn. Lett. **5**, 85–100 (2020). https://doi.org/10.21809/rilemtechlett.2020.127
16. Backus, J., McPolin, D., Basheer, M., Long, A., Holmes, N.: Exposure of mortars to cyclic chloride ingress and carbonation. Adv. Cem. Res. **25**, 3–11 (2013). https://doi.org/10.1680/adcr.12.00029
17. Banthia, N., Bhargava, A.: Permeability of stressed concrete and role of fiber reinforcement. ACI Mater. J. **104**, 70–77 (2007)
18. Bone, J.R., Stafford, R., Hall, A.E., Herbert, R.J.H.: Biodeterioration and bioprotection of concrete assets in the coastal environment. Int. Biodeterior. Biodegrad. **175**, 105507 (2022). https://doi.org/10.1016/j.ibiod.2022.105507
19. Bourbatache, K., Millet, O., Aït-Mokhtar, A., Amiri, O.: Modeling the chlorides transport in cementitious materials by periodic homogenization. Transp. Porous Media **94**, 437–459 (2012). https://doi.org/10.1007/s11242-012-0013-1
20. Buckley, L.J., Carter, M.A., Wilson, M.A., Scantlebury, J.D.: Methods of obtaining pore solution from cement pastes and mortars for chloride analysis. Cem. Concr. Res. **37**, 1544–1550 (2007). https://doi.org/10.1016/j.cemconres.2007.08.009
21. Bérubé, M.A., Carles-Gibergues, A.: La durabilité des bétons face à l'alcali réaction (The durability of concretes face to alkali reactions). In: La Durabilité des Bétons (The Durability

of Concretes), pp. 285–350. Presses de l'Ecole Nationales des Ponts et Chaussée, Paris (1992). (in French)
22. Castel, A., François, R., Arliguie, G.: Effect of loading on carbonation penetration in reinforced concrete elements. Cem. Concr. Res. **29**, 561–565 (1999)
23. Castro, P., De Rincon, O.T., Pazini, E.J.: Interpretation of chloride profiles from concrete exposed to tropical marine environments. Cem. Concr. Res. **31**, 529–537 (2001). https://doi.org/10.1016/S0008-8846(01)00453-7
24. Castro-Borges, P., Torres-Acosta, A.A., Balancán-Zapata, M.G.: Long term correlation between concrete cracking and corrosion in natural marine micro-environments. Mater. Struct. **54**, 233 (2021). https://doi.org/10.1617/s11527-021-01821-8
25. Castro-Borges, P., Cabrera-Madrid, J. A., Balancán-Zapata, M.G., Torres-Acosta, A.A.: Concrete cracking in two marine micro-climates. In: 2nd R.N. Raikar International Conference and Banthia-Basheer International Symposium on Advances in Science and Technology of Concrete, 18–19 December 2015, Mumbai, India, pp. 146–156 (2015)
26. Chalee, W., Jaturapitakkul, C.: Effects of w/b ratios and fly ash finenesses on chloride diffusion coefficient of concrete in marine environment. Mater. Struct. **42**, 505–514 (2008). https://doi.org/10.1617/s11527-008-9398-2
27. Chalee, W., Sasakul, T., Suwanmaneechot, P., Jaturapitakkul, C.: Utilization of rice husk–bark ash to improve the corrosion resistance of concrete under 5-year exposure in a marine environment. Cem. Concr. Compos. **37**, 47–53 (2013). https://doi.org/10.1016/j.cemconcomp.2012.12.007
28. Chang, H.L., Jin, Z.Q., Zhao, T.J., Wang, B.Z., Li, Z., Liu, J.: Capillary suction induced water absorption and chloride transport in non-saturated concrete: the influence of humidity, mineral admixtures and sulfate ions. Constr. Build. Mater. **236**, 117581 (2020). https://doi.org/10.1016/j.conbuildmat.2019.117581
29. Chatterji, S.: Mechanism of the $CaCl_2$ attack on Portland cement concrete. Cem. Concr. Res. **8**, 461–467 (1978). https://doi.org/10.1016/0008-8846(78)90026-1
30. Cheewaket, T., Jaturapitakkul, C., Chalee, W.: Concrete durability presented by acceptable chloride level and chloride diffusion coefficient in concrete: 10-year results in marine site. Mater. Struct. **47**, 1501–1511 (2014). https://doi.org/10.1617/s11527-013-0131-4
31. Cherif, R., Hamami, A.E.A., Aït-Mokhtar, A.: Effects of leaching and chloride migration on the microstructure and pore solution of blended cement pastes during a migration test. Constr. Build. Mater. **240**, 117934 (2020). https://doi.org/10.1016/j.conbuildmat.2019.117934
32. Cidreira, K.G., Sanchez, T., Conciatori, D., Chouinard, L.: Monitoring environmental and climatic exposure conditions for structures in cold regions. J. Cold Reg. Eng. **35**, 04021007 (2021). https://doi.org/10.1061/(ASCE)CR.1943-5495.0000249
33. Cwirzen, A., Sztermen, P., Habermehl-Cwirzen, K.: Effect of Baltic seawater and binder type on frost durability of concrete. J. Mater. Civ. Eng. **26**, 283–287 (2014). https://doi.org/10.1061/(ASCE)MT.1943-5533.0000803
34. Danner, T., Hjorth Jakobsen, U., Geiker, M.R.: Mineralogical sequence of self-healing products in cracked marine concrete. Minerals **9**, 284 (2019). https://doi.org/10.3390/min9050284
35. Delagrave, A., Marchand, J., Ollivier, J.-P., Julien, S., Hazrati, K.: Chloride binding capacity of various hydrated cement paste systems. Adv. Cem. Based Mater. **6**, 28–35 (1997). https://doi.org/10.1016/S1065-7355(97)90003-1
36. Dhir, R.K., El-Mohr, M.A.K., Dyer, T.D.: Chloride binding in GGBS concrete. Cem. Concr. Res. **26**, 1767–1773 (1996). https://doi.org/10.1016/S0008-8846(96)00180-9
37. Dhir, R.K., Hewlett, P.C., Chan, Y.N.: Near-surface characteristics of concrete: abrasion resistance. Mater. Struct. **24**, 122–128 (1991). https://doi.org/10.1007/BF02472473
38. Duchesne, J., Bertron, A.: Performance of cement-based materials in aggressive aqueous environments state-of-the-art report RILEM TC 211—PAE Leaching of Cementitious Materials by Pure Water and Strong Acids (HCl and $HNO_3$), pp. 91–112. Springer, Netherlands, Dordrecht (2012)

39. Edvardsen, C.: Water permeability and autogenous healing of cracks in concrete. ACI Mater. J. **96**(4), 448–454 (1999)
40. Fagerlund, G.: On the service life of concrete exposed to frost action. In: Proceedings of International Workshop in the Resistance of Concrete to Scaling due to Freezing in the Presence of Deicing Salts, pp. 23–41. E & FN Spon, London, UK (1997)
41. Fagerlund, G.: Mechanical damage and fatigue effects associated with freeze-thaw of materials. In: Proceedings of International RILEM Workshop of Frost Resistance of Concrete, pp. 117–132. RILEM Publications SARL, Paris, France (2002)
42. Fan, Z., Yang, F., Huang, J., Wang, S.: Experimental study on long- term exposure test of marine concrete. Port Waterw. Eng. **30**(9), 45–48+57 (2005). (in Chinese)
43. Fjendbo, S., Sørensen, H.E., De Weerdt, K., Geiker, M.R.: The square root method for chloride ingress prediction—applicability and limitations. Mater. Struct. **54**, 1–19 (2021)
44. Fjendbo, S., Sørensen, H.E., De Weerdt, K., Jakobsen, U.H., Geiker, M.R.: Correlating the development of chloride profiles and microstructural changes in marine concrete up to ten years. Cem. Concr. Compos. **131**, 104590 (2022). https://doi.org/10.1016/j.cemconcomp.2022.104590
45. Fukute, T., Hamada, H.: A study on the durability of concrete exposed in marine environment for 20 years. Report of the Port and Harbour Research Institute, vol. 31(5). Ministry of Transport, Yokosuka, Japan (1993). (in Japanese)
46. Furcas, F.E., Lothenbach, B., Isgor, O.B., Mundra, S., Zhang, Z., Angst, U.M.: Solubility and speciation of iron in cementitious systems. Cem. Concr. Res. **151**, 106620 (2022). https://doi.org/10.1016/j.cemconres.2021.106620
47. Gaylarde, C.C., Ortega-Morales, B.O.: Biodeterioration and chemical corrosion of concrete in the marine environment: too complex for prediction. Microorganisms **11**, 2438 (2023). https://doi.org/10.3390/microorganisms11102438
48. Geiker, M.R., Laugesen, P.: On the effect of laboratory conditioning and freeze/thaw exposure on moisture profiles in HPC. Cem. Concr. Res. **31**, 1831–1836 (2001). https://doi.org/10.1016/S0008-8846(01)00643-3
49. Geiker, M.R.: State-of-the-Art report Fly ash in concrete, Danish experience. Statens vegvesens rapporter 370, Norwegian Public Roads Administration, Oslo, Norway (2015)
50. Glass, G.K., Page, C.L., Short, N.R.: Factors affecting the corrosion rate of steel in carbonated mortars. Corros. Sci. **32**(12), 1283–1294 (1991)
51. Glasser, F.P., Pedersen, J., Goldthorpe, K., Atkins, M.: Solubility reactions of cement components with NaCl solutions: I. $Ca(OH)_2$ and C-S-H. Adv. Cem. Res. **17**, 57–64 (2005). https://doi.org/10.1680/adcr.2005.17.2.57
52. Gong, F., Jacobsen, S.: Modeling of water transport in highly saturated concrete with wet surface during freeze/thaw. Cem. Concr. Res. **115**, 294–307 (2019). https://doi.org/10.1016/j.cemconres.2018.08.013
53. Gong, F., Wang, Y., Zhang, D., Ueda, T.: Mesoscale simulation of deformation for mortar and concrete under cyclic freezing and thawing stress. J. Adv. Concr. Technol. **13**, 291–304 (2015). https://doi.org/10.3151/jact.13.291
54. Gong, F., Zhi, D., Jia, J., Wang, Z., Ning, Y., Zhang, B., Ueda, T.: Data-based statistical analysis of laboratory experiments on concrete frost damage and its implications on service life prediction. Materials **15**(18) 6282 (2022). https://doi.org/10.3390/ma15186282
55. González, J.A., Algaba, S., Andrade, C.: Corrosion of reinforcing bars in carbonated concrete. Br. Corros. J. **3**, 135–139 (1980)
56. González, J.A., Andrade, C.: Effect of carbonation, chlorides and relative ambient humidity on the corrosion of galvanized rebars embedded in concrete. Br. Corros. J. **17**(1), 21–28 (1982)
57. Hasan, M., Okuyama, H., Sato, Y., Ueda, T.: Stress-strain model of concrete damaged by freezing and thawing cycles. J. Adv. Concr. Technol. **2**, 89–99 (2004). https://doi.org/10.3151/jact.2.89
58. Hashin, Z.: Analysis of Composite Materials—A Survey. J. Appl. Mech. **50**, 481–505 (1983). https://doi.org/10.1115/1.3167081

59. Hemstad, P., Machner, A., De Weerdt, K.: The effect of artificial leaching with HCl on chloride binding in ordinary Portland cement paste. Cem. Concr. Res. **130**, 105976 (2020). https://doi.org/10.1016/j.cemconres.2020.105976
60. Hooton, D., Geiker, M., Bentz, E.: Effects of curing on chloride ingress and implications on service life. ACI Mater. J. **99**, 201–206 (2002)
61. Hornbostel, K., Elsener, B., Angst, U.M., Larsen, C.K., Geiker, M.R.: Limitations of the use of concrete bulk resistivity as an indicator for the rate of chloride-induced macro-cell corrosion. Struct. Concr. **18**, 326–333 (2017). https://doi.org/10.1002/suco.201500141
62. Huang, J., Fan, Z., Wang, S., Pan, D.: Analysis of exposure experiment and application of coatings for marine concrete structures. China Harb. Eng. **31**(2), 16–19+54 (2006). (in Chinese)
63. Isgor, O.B., Weiss, W.J.: A nearly self-sufficient framework for modelling reactive-transport processes in concrete. Mater. Struct. **52**, 3 (2018). https://doi.org/10.1617/s11527-018-1305-x
64. Itoh, Y., Yoshida, A., Tsuchiya, M., Katoh, K., Sasaki, K., Saeki, H.: An experimental study on abrasion of concrete due to sea ice. In: Offshore Technology Conference, 2–5 May 1988, Huston, Texas, OnePetro, Paper No. OTC-5687-MS (1988). https://doi.org/10.4043/5687-MS
65. Jacobsen, S., Scherer, G.W., Schulson, E.M.: Concrete–ice abrasion mechanics. Cem. Concr. Res. **73**, 79–95 (2015). https://doi.org/10.1016/j.cemconres.2015.01.001
66. Jacobsen, S., Sellevold, E.J., Matala, S.: Frost durability of high strength concrete: effect of internal cracking on ice formation. Cem. Concr. Res. **26**, 919–931 (1996). https://doi.org/10.1016/0008-8846(96)00066-X
67. Jakobsen, U.H., De Weerdt, K., Geiker, M.R.: Elemental zonation in marine concrete. Cem. Concr. Res. **85**, 12–27 (2016). https://doi.org/10.1016/j.cemconres.2016.02.006
68. Jayakumar, S., Saravanane, R.: Biodeterioration of coastal concrete structures by marine green algae. Int. J. Civ. Eng. **8**, 352–361 (2010)
69. Johannesson, B.: Dimensional and ice content changes of hardened concrete at different freezing and thawing temperatures. Cem. Concr. Compos. **32**, 73–83 (2010). https://doi.org/10.1016/j.cemconcomp.2009.09.001
70. Kanavaris, F., Coelho, M., Mesquita Ferreira, N., Azenha, M., Andrade, C.: A review on the effects of cracking and crack width on corrosion of reinforcement in concrete. Struct. Concr. **24**, 7272–7294 (2023). https://doi.org/10.1002/suco.202300227
71. Karthick, S.P., Madhavamayandi, A., Muralidharan, S., Saraswathy, V.: Electrochemical process to improve the durability of concrete structures. J. Build. Eng. **7**, 273–280 (2016). https://doi.org/10.1016/j.jobe.2016.07.005
72. Kayyali, O.A., Haque, M.N.: Chloride penetration and the ratio of $C\ell-/OH-$ in the pores of cement paste. Cem. Concr. Res. **18**, 895–900 (1988). https://doi.org/10.1016/0008-8846(88)90025-7
73. Kjeldsen, A.M., Geiker, M.R.: On the interpretation of low temperature calorimetry data. Mater. Struct. **41**(1), 213–224 (2007). https://doi.org/10.1617/s11527-007-9239-8
74. Koleva, D.A., Hu, J., Fraaij, A.L.A., van Breugel, K., de Wit, J.H.W.: Microstructural analysis of plain and reinforced mortars under chloride-induced deterioration. Cem. Concr. Res. **37**, 604–617 (2007). https://doi.org/10.1016/j.cemconres.2006.12.001
75. Kwon, S.-J., Lee, H.-S., Karthick, S., Saraswathy, V., Yang, H.-M.: Long-term corrosion performance of blended cement concrete in the marine environment – a real-time study. Constr. Build. Mater. **154**, 349–360 (2017). https://doi.org/10.1016/j.conbuildmat.2017.07.237
76. Kyle, A., Bergsma, B.P., Hansson, C.M.: Chloride concentration in the pore solution of Portland cement paste and Portland cement concrete. Cem. Concr. Res. **63**, 35–37 (2014). https://doi.org/10.1016/j.cemconres.2014.04.008
77. Laplante, P., Aitcin, P.-C., Vézina, D.: Abrasion resistance of concrete. J. Mater. Civ. Eng. **3**, 19–28 (1991). https://doi.org/10.1061/(ASCE)0899-1561(1991)3:1(19)
78. Li, K.F.: Durability Design of Concrete Structures: Phenomena, Modeling, and Practice. Wiley, London (2016)
79. Li, K.F., Li, C.Q.: Modeling hydroionic transport in cement-based porous materials under drying-wetting actions. J. Appl. Mech. **80**(2), 020904 (2013). https://doi.org/10.1115/1.4007907

80. Li, L., Li, K.F.: Permeability of microcracked solids with random crack networks: role of connectivity and opening aperture. Transp. Porous Media **109**, 217–237 (2015). https://doi.org/10.1007/s11242-015-0510-0
81. Li, K.F., Li, L.: Crack-altered durability properties and performance of structural concretes. Cem. Concr. Res. **124**, 105811 (2019). https://doi.org/10.1016/j.cemconres.2019.105811
82. Li, L., Li, K.F.: Permeability of microcracked porous solids: modeling from homogenization to percolation. J. Eng. Mech. ASCE **148**, 04022045 (2022). https://doi.org/10.1061/(ASCE)EM.1943-7889.0002132
83. Li, L., Liu, Q., Tang, L., Hu, Z., Wen, Y., Zhang, P.: Chloride penetration in freeze–thaw induced cracking concrete: a numerical study. Constr. Build. Mater. **302**, 124291 (2021). https://doi.org/10.1016/j.conbuildmat.2021.124291
84. Li, K.F., Ma, M.J., Wang, X.M.: Experimental study of water flow behaviour in narrow fractures of cementitious materials. Cem. Concr. Compos. **33**, 1009–1013 (2011). https://doi.org/10.1016/j.cemconcomp.2011.08.005
85. Li, W., Pour-Ghaz, M., Castro, J., Weiss, W.: Water absorption and critical degree of saturation relating to freeze-thaw damage in concrete pavement joints. J. Mater. Civ. Eng. **24**, 299–307 (2012). https://doi.org/10.1061/(ASCE)MT.1943-5533.0000383
86. Lim, C.C., Gowripalan, N., Sirivivatnanon, V.: Microcracking and chloride permeability of concrete under uniaxial compression. Cement Concr. Compos. **22**, 353–360 (2000)
87. Lindvall, A.: Chloride ingress data from field and laboratory exposure – influence of salinity and temperature. Cem. Concr. Compos. **29**, 88–93 (2007). https://doi.org/10.1016/j.cemconcomp.2006.08.004
88. Liu, J., Qiu, Q., Chen, X., Xing, F., Han, N., He, Y., Ma, Y.: Understanding the interacted mechanism between carbonation and chloride aerosol attack in ordinary Portland cement concrete. Cem. Concr. Res. **95**, 217–225 (2017). https://doi.org/10.1016/j.cemconres.2017.02.032
89. Liu, Y.-W., Yen, T., Hsu, T.-H.: Abrasion erosion of concrete by water-borne sand. Cem. Concr. Res. **36**, 1814–1820 (2006). https://doi.org/10.1016/j.cemconres.2005.03.018
90. Løset, S., Shkhinek, K., Høyland, K.: Ice Physics and Mechanics. Norwegian University of Science and Technology, Trondheim, 125p. (1998)
91. Machner, A., Bjørndal, M.H., Justnes, H., Hanžič, L., Šajna, A., Gu, Y., Bary, B., Ben Haha, M., Geiker, M.R., De Weerdt, K.: Effect of leaching on the composition of hydration phases during chloride exposure of mortar. Cem. Concr. Res. **153**, 106691 (2022). https://doi.org/10.1016/j.cemconres.2021.106691
92. Marcos-Meson, V., Geiker, M., Fischer, G., Solgaard, A., Jakobsen, U.H., Danner, T., Edvardsen, C., Skovhus, T.L., Michel, A.: Durability of cracked SFRC exposed to wet-dry cycles of chlorides and carbon dioxide – multiscale deterioration phenomena. Cem. Concr. Res. **135**, 106120 (2020). https://doi.org/10.1016/j.cemconres.2020.106120
93. Maus, S., Bahafid, S., Hendriks, M., Jacobsen, S., Geiker, M.R.: X-ray micro-tomographic imaging and modelling of saline ice properties in concrete frost salt scaling experiments. Cold Reg. Sci. Technol. **208**, 103780 (2023). https://doi.org/10.1016/j.coldregions.2023.103780
94. Mays, G.C. (ed.): Durability of Concrete Structures, Investigation, Repair, Protection, 1st edn. CRC Press, London (1991)
95. Mazer, W., Lima, M., Medeiros-Junior, R., Wickzick, L.: Chloride ingress into concrete under different conditions of temperature and marine zones. Mag. Concr. Res. **73**, 1180–1188 (2020). https://doi.org/10.1680/jmacr.19.00451
96. Mehta, P.K., Burrows, R.W.: Building durable structures in the 21st century. Concr. Int. **3**, 57–63 (2001)
97. Mehta, P.K.: Concrete durability-fifty years progress? In: SP-126: Durability of Concrete: Second International Conference, Montreal, Canada, American Concrete Institute, pp. 1–32 (1991)
98. Meira, G.R., Padaratz, I.J., Alonso, C., Andrade, C.: Effect of distance from sea on chloride aggressiveness in concrete structures in Brazilian coastal site. Mater. Constr. **53**, 179–188 (2003). https://doi.org/10.3989/mc.2003.v53.i271-272.302

99. Melchers, R.E., Li, C.Q.: Phenomenological modeling of reinforcement corrosion in marine environments. ACI Mater. J. **103**(1), 25–32 (2006)
100. Mesbah, A., François, M., Cau-dit-Coumes, C., Frizon, F., Filinchuk, Y., Leroux, F., Ravaux, J., Renaudin, G.: Crystal structure of Kuzel's salt 3CaO·Al2O3·1/2CaSO4·1/2CaCl2·11H2O determined by synchrotron powder diffraction. Cem. Concr. Res. **41**, 504–509 (2011). https://doi.org/10.1016/j.cemconres.2011.01.015
101. Meziani, H., Skoczylas, F.: An experimental study of the mechanical behaviour of a mortar and of its permeability under deviatoric loading. Mater. Struct. **32**, 403–409 (1999)
102. Michael, H., Scott, K., Koneshloo, M., Yu, X., Khan, M., Li, K.: Geologic influence on groundwater salinity drives large seawater circulation through the continental shelf: geology drives seawater circulation. Geophys. Res. Lett. **43**, 10–782 (2016). https://doi.org/10.1002/2016GL070863
103. Midgley, H.G., Illston, J.M.: The penetration of chlorides into hardened cement pastes. Cem. Concr. Res. **14**, 546–558 (1984). https://doi.org/10.1016/0008-8846(84)90132-7
104. Mohammed, T., Hamada, H., Yamaji, T.: Concrete after 30 years of exposure - part 1: mineralogy, microstructures, and interfaces. ACI Mater. J. **101**, 3–12 (2004)
105. Mohammed, T., Yamaji, T., Aoyama, T., Hamada, H.: Marine durability of 15-year old uncracked and precracked concrete made with different cements. Doboku Gakkai Ronbunshu (Proc. JSCE) **2002**(697), 201–214 (2002). https://doi.org/10.2208/jscej.2002.697_201
106. Mohammed, T. U., Otsuki, N., Hisada, M., Hamada, H.: Marine durability of 23-year-old reinforced concrete beams. In: Malholtra, V.M. (ed.) Durability of Concrete, Proceedings of the fifth CANMET/ACI International Conference, SP-192, vol. 2, pp. 1071–1088. American Concrete Institute, Farmington Hills
107. Mohammed, T., Yamaji, T., A., T., Hamada, H.: Marine Durability of 15-Year Old Concrete Specimens Made with Ordinary Portland, Slag and Fly Ash Cement. ACI Spec. Publ. SP **199**(2), 541–560 (2001)
108. Moore, A., Beushausen, H., Otieno, M., Ndawula, J., Alexander, M.: Oxygen availability and corrosion propagation in RC structures in the marine environment—inferences from field and laboratory studies. Corros. Mater. Degrad. **3**, 363–375 (2022). https://doi.org/10.3390/cmd3030022
109. Moragues, A., Macias, A., Andrade, C.: Equilibrium of the chemical composition of the concrete pore solution. Part I: comparative study of synthetic and extracted solutions. Cem. Concr. Res. **17**, 173–182 (1987)
110. Moragues, A., Macias, A., Andrade, C., Losada, J.: Equilibria of the chemical composition of the pore concrete solution. Part II: calculation of the equilibria constants of the synthetic solutions. Cem. Concr. Res. **18**, 342–350 (1988)
111. Morcillo, M., Chico, B., Mariaca, L., Otero, E.: Salinity in marine atmospheric corrosion: its dependence on the wind regime existing in the site. Corros. Sci. **42**(1), 91–104 (2000)
112. Møen, E., Jacobsen, S., Myhra, H.: Ice abrasion data on concrete structures – an overview, State of the art. SINTEF report SBF BK A07036, SINTEF Building and Infrastructure, Trondheim, Norway (2007)
113. Ngala, V.T., Page, C.L.: Effects of carbonation on pore structure and diffusional properties of hydrated cement pastes. Cem. Concr. Res. **27**, 995–1007 (1997). https://doi.org/10.1016/S0008-8846(97)00102-6
114. Nielsen, E.P., Geiker, M.R.: Chloride diffusion in partially saturated cementitious material. Cem. Concr. Res. **33**(1), 133–138 (2003). https://doi.org/10.1016/S0008-8846(02)00939-0
115. Page, C.L., Short, N.R., El Tarras, A.: Diffusion of chloride ions in hardened cement pastes. Cem. Concr. Res. **11**, 395–406 (1981). https://doi.org/10.1016/0008-8846(81)90111-3
116. Poulsen, S.L., Sørensen, H.E.: Chloride ingress in old Danish bridges. In: Proceedings of 2nd International Congress on Durability of Concrete (ICDC), 4–6 December 2014, New Delhi, India (2014)
117. Pratiwi, W., Putra, F., Triwulan, T., Tajunnisa, Y., Husin, N., Wulandari, K.: A review of concrete durability in marine environment. IOP Conf. Ser. Mater. Sci. Eng. **1175**, 012018 (2021). https://doi.org/10.1088/1757-899X/1175/1/012018

118. Pu, Q., Jiang, L., Xu, J., Chu, H., Xu, Y., Zhang, Y.: Evolution of pH and chemical composition of pore solution in carbonated concrete. Constr. Build. Mater. **28**, 519–524 (2012). https://doi.org/10.1016/j.conbuildmat.2011.09.006
119. Qiao, C., Suraneni, P., Weiss, J.: Damage in cement pastes exposed to NaCl solutions. Constr. Build. Mater. **171**, 120–127 (2018). https://doi.org/10.1016/j.conbuildmat.2018.03.123
120. Ramachandran, V.S.: Possible states of chloride in the hydration of tricalcium silicate in the presence of calcium chloride. Mater. Struct. **19**, 3–12 (1971)
121. Ramezanianpour, A.A., Pourkhorshidi, A.R., Sobhani, J., Moodi, F.: Durability of concrete containing blended cements in harsh marine environments: 18 years exposure study. Constr. Build. Mater. **299**, 123863 (2021). https://doi.org/10.1016/j.conbuildmat.2021.123863
122. Regourd, M. Hornain, H., Montureux, B.: Microstructure of concrete in aggressive environments. In: Sereda P.J., Litvan, G.G. (eds.) Durability of Building Materials and Components. ASTM International, US (1980). https://doi.org/10.1520/STP36064S
123. Reinhardt, H.W.: Penetration and permeability of concrete: barriers to organic and contaminating liquids. RILEM report 16, E & FN Spon, London, UK (1997)
124. Revert, A.B., Hornbostel, K., De Weerdt, K., Geiker, M.R.: Macrocell corrosion in carbonated Portland and Portland-fly ash concrete - contribution and mechanism. Cem. Concr. Res. **116**, 273–283 (2019). https://doi.org/10.1016/j.cemconres.2018.12.005
125. Saeki, T., Ohga, H., Nagataki, S.: Change in micro-structure of concrete due to carbonation. Doboku Gakkai Ronbunshu (Proc. JSCE) **1990**(420), 33–42 (1990)
126. Saillio, M., Baroghel-Bouny, V., Barberon, F.: Chloride binding in sound and carbonated cementitious materials with various types of binder. Constr. Build. Mater. **68**, 82–91 (2014). https://doi.org/10.1016/j.conbuildmat.2014.05.049
127. Sanchez-Silva, M., Rosowsky, D.V.: Biodeterioration of construction materials: state of the art and future challenges. J. Mater. Civ. Eng. **20**, 352–365 (2008). https://doi.org/10.1061/(ASCE)0899-1561(2008)20:5(352)
128. Sandberg, P., Tang, L., Andersen, A.: Recurrent studies of chloride ingress in uncracked marine concrete at various exposure times and elevations. Cem. Concr. Res. **28**, 1489–1503 (1998). https://doi.org/10.1016/S0008-8846(98)00124-0
129. Schueremans, L., Van Gemert, D., Giessler, S.: Chloride penetration in RC-structures in marine environment – long term assessment of a preventive hydrophobic treatment. Constr. Build. Mater. **21**, 1238–1249 (2007). https://doi.org/10.1016/j.conbuildmat.2006.05.006
130. Scott, A., Alexander, M.G.: The influence of binder type, cracking and cover on corrosion rates of steel in chloride-contaminated concrete. Mag. Concr. Res. **59**, 495–505 (2007)
131. Scott, B., Safiuddin, M.: Abrasion resistance of concrete – design, construction and case study. Concr. Res. Lett. **6**, 136–148 (2015)
132. Setzer, M.J.: Micro-ice-lens formation in porous solid. J. Colloid Interface Sci. **243**, 193–201 (2001). https://doi.org/10.1006/jcis.2001.7828
133. Shafiro, B., Kachanov, M.: Anisotropic effective conductivity of materials with nonrandomly oriented inclusions of diverse ellipsoidal shapes. J. Appl. Phys. **87**, 8561–8569 (2000). https://doi.org/10.1063/1.373579
134. Shamsutdinova, G., Hendriks, M.A.N., Jacobsen, S.: Concrete-ice abrasion: wear, coefficient of friction and ice consumption. Wear **416–417**, 27–35 (2018). https://doi.org/10.1016/j.wear.2018.09.007
135. Shang, H.S., Song, Y.P.: Experimental study of strength and deformation of plain concrete under biaxial compression after freezing and thawing cycles. Cem. Concr. Res. **36**, 1857–1864 (2006). https://doi.org/10.1016/j.cemconres.2006.05.018
136. Stewart, M.G., Al-Harthy, A.: Pitting corrosion and structural reliability of corroding RC structures: experimental data and probabilistic analysis. Reliab. Eng. Syst. Saf. **93**, 373–382 (2008). https://doi.org/10.1016/j.ress.2006.12.013
137. Suryavanshi, A.K., Scantlebury, J.D., Lyon, S.B.: Mechanism of Friedel's salt formation in cements rich in tri-calcium aluminate. Cem. Concr. Res. **26**, 717–727 (1996). https://doi.org/10.1016/S0008-8846(96)85009-5

138. Sutter, L., Peterson, K., Touton, S., Van Dam, T., Johnston, D.: Petrographic evidence of calcium oxychloride formation in mortars exposed to magnesium chloride solution. Cem. Concr. Res. **36**(8), 1533–1541 (2006). https://doi.org/10.1016/j.cemconres.2006.05.022
139. Sørensen, H., Jönsson, U., Munch-Petersen, C.: Chloride penetration into concrete – Comparison of results from field exposure tests and laboratory tests. In: Proceedings of 2nd International Congress on Durability of Concrete (ICDC), 4–6 December 2014, New Delhi, India (2014). https://doi.org/10.13140/RG.2.1.4088.2960
140. Thomas, M.D.A., Bamforth, P.B.: Modelling chloride diffusion in concrete: effect of fly ash and slag. Cem. Concr. Res. **29**, 487–495 (1999). https://doi.org/10.1016/S0008-8846(98)00192-6
141. Thomas, M., Bremner, T.: Performance of lightweight aggregate concrete containing slag after 25 years in a harsh marine environment. Cem. Concr. Res. **42**, 358–364 (2012). https://doi.org/10.1016/j.cemconres.2011.10.009
142. Thomas, M.D.A., Matthews, J.D.: Performance of pfa concrete in a marine environment—10-year results. Cem. Concr. Compos. **26**, 5–20 (2004). https://doi.org/10.1016/S0958-9465(02)00117-8
143. Torres-Acosta, A., Castro-Borges, P.: Corrosion-induced cracking of concrete elements exposed to a natural marine environment for five years. Corrosion **69**, 1122–1131 (2013). https://doi.org/10.5006/0844
144. Tran, V.Q., Soive, A., Baroghel-Bouny, V.: Modelisation of chloride reactive transport in concrete including thermodynamic equilibrium, kinetic control and surface complexation. Cem. Concr. Res. **110**, 70–85 (2018). https://doi.org/10.1016/j.cemconres.2018.05.007
145. Tuutti, K.: Corrosion of Steel in Concrete, 468p. Swedish Cement and Concrete Research Institute, Stockholm (1982)
146. Wang, D., Ma, Y., Kang, M., Ju, Y., Zeng, C.: Durability of reactive powder concrete containing mineral admixtures in seawater erosion environment. Constr. Build. Mater. **306**, 124863 (2021). https://doi.org/10.1016/j.conbuildmat.2021.124863
147. De Weerdt, K., Colombo, A., Coppola, L., Justnes, H., Geiker, M.R.: Impact of the associated cation on chloride binding of Portland cement paste. Cem. Concr. Res. **68**, 196–202 (2015). https://doi.org/10.1016/j.cemconres.2014.01.027
148. De Weerdt, K., Orsáková, D., Müller, A.C.A., Larsen, C.K., Pedersen, B., Geiker, M.R.: Towards the understanding of chloride profiles in marine exposed concrete, impact of leaching and moisture content. Constr. Build. Mater. **120**, 418–431 (2016). https://doi.org/10.1016/j.conbuildmat.2016.05.069
149. Wei, S., Zhenglong, J., Liu, H., Zhou, D., Sanchez-Silva, M.: Microbiologically induced deterioration of concrete - a review. Braz. J. Microbiol. **44**, 1001–1007 (2013). https://doi.org/10.1590/S1517-83822014005000006
150. Xu, Y.: The influence of sulphates on chloride binding and pore solution chemistry. Cem. Concr. Res. **27**, 1841–1850 (1997). https://doi.org/10.1016/S0008-8846(97)00196-8
151. Yang, Z., Weiss, W., Olek, J.: Water transport in concrete damaged by tensile loading and freeze–thaw cycling. J. Mater. Civ. Eng. ASCE **18**, 424–434 (2006). https://doi.org/10.1061/(ASCE)0899-1561(2006)18:3(424)
152. Yao Y., Wang L., Wittmann F.H.: Publications on Durability of Reinforced Concrete Structures under Combined Mechanical Loads and Environmental Actions: An Annotated Bibliography, Aedificatio Pulishers, Feriburg, Germany (2013)
153. Ye, H., Jin, X., Fu, C., Jin, N., Xu, Y., Huang, T.: Chloride penetration in concrete exposed to cyclic drying-wetting and carbonation. Constr. Build. Mater. **112**, 457–463 (2016). https://doi.org/10.1016/j.conbuildmat.2016.02.194
154. Zhou, C.S., Li, K.F.: Transport properties of concrete altered by crack-induced damage. J. Mater. Civ. Eng. ASCE **27**, A4014001 (2015). https://doi.org/10.1061/(ASCE)MT.1943-5533.0000983
155. Zhou, C.S., Li, K.F., Han, J.G.: Characterizing the effect of compressive damage on transport properties of cracked concretes. Mater. Struct. **45**, 381–392 (2012). https://doi.org/10.1617/s11527-011-9771-4

156. Zhou, C.S., Li, K.F., Pang, X.Y.: Effect of crack density and connectivity on the permeability of microcracked solids. Mech. Mater. **43**, 969–978 (2011). https://doi.org/10.1016/j.mechmat.2011.08.011
157. Zhou, C.S., Li, K.F., Pang, X.Y.: Geometry of crack network and its impact on transport properties of concrete. Cem. Concr. Res. **42**, 1261–1272 (2012). https://doi.org/10.1016/j.cemconres.2012.05.017
158. Zhu, W.J., François, R., Zhang, C.P., Zhang, D.L.: Propagation of corrosion-induced cracks of the RC beam exposed to marine environment under sustained load for a period of 26 years. Cem. Concr. Res. **103**, 66–76 (2008). https://doi.org/10.1016/j.cemconres.2017.09.014

# Chapter 3
# Long-Term Field Exposure Data: Information and Presentation

**Junjie Zeng, Haicheng Yang, Pedro Castro Borges, Kefei Li, Quanwang Li, Bowei Li, Mette Rica Geiker, Riccardo Maddalena, Carmen Andrade, Luping Tang, José Fernando Martirena-Hernández, Peng Zhang, and Véronique Baroghel-Bouny**

J. Zeng (✉) · H. Yang
CCCC Fourth Harbor Engineering Co., Ltd., Guangzhou, China
e-mail: junjie.zeng2020@foxmail.com

P. C. Borges
Centro de Investigación y de Estudios Avanzados del IPN, Unidad Mérida, Mérida, Yucatán, México

K. Li · Q. Li · B. Li
Civil Engineering Department, Tsinghua University, Beijing, China

M. R. Geiker
Department of Structural Engineering, Norwegian University of Science and Technology, Trondheim, Norway

R. Maddalena
School of Engineering, Cardiff University, Cardiff, UK

C. Andrade
International Center for Numerical Methods in Engineering (CIMNE), Barcelona, Spain

L. Tang
Department of Architecture and Civil Engineering, Chalmers University of Technology, Gothenburg, Sweden

J. F. Martirena-Hernández
Facultad de Construcciones, Universidad Central de las Villas, Santa Clara, Villa Clara, Cuba

P. Zhang
Qingdao University of Technology, Qingdao, China

V. Baroghel-Bouny
Université Gustave Eiffel, Champs sur Marne, France

 
K. Li et al. (eds.), *Long-Term Performance of Structural Concretes in Marine Exposure Conditions*, RILEM State-of-the-Art Reports Volume,
https://doi.org/10.1007/978-3-032-06959-7_3

## 3.1 Marine Exposure Sites: Characteristics

The original purpose of exposure sites is to accommodate concrete specimens and elements to observe their in-field performance in real marine environments. Thus, all the characteristics related to the exposure sites constitute the very first information for long-term exposure practice, including both the physical conditions provided by the exposure sites and the technical support for long-term observation provided by the operators or the owners. From the available experiences so far [13, 39], the main characteristics of exposure sites can include but are not limited by the following items:

(a) the geographical location and orientation of exposure sites and facilities,
(b) the functional planning of exposure area, positioning for different types of specimens and retrieval schemes of exposed specimens,
(c) the meteorology of surrounding or regional atmosphere, such as temperature, relative humidity, $CO_2$ concentration, time of wetness and chloride deposition rate,
(d) the characteristics and classification of atmospheric corrosivity for the exposure sites and facilities [21],
(e) the hydrology of seawater such as waves and tides, temperature, salinity, pH value and ion concentrations,
(f) the biological growth rates (if any), and
(g) the methods to collect, record and communicate the data relevant to exposure sites.

Figure 3.1 shows a typical marine exposure site situated on the coastal area. This exposure site can be characterized by such information as the distance from the coastline, the location of seawater, the distribution of buildings, the placement of exposed specimens, and accessibility to transport such as haulage roads.

Figure 3.2 shows the collecting of seawater hydrology information, in Ling'ding Ocean, China, and the recording of meteorological condition on the exposure station established on the Hong Kong-Zhuhai-Macau (HZM) sea link project site. Through these measurements the crucial environmental factors, such as salinity of seawater and atmospheric humidity and temperature, can be recorded and monitored.

Among the exposure sites one can differentiate between unshielded sites and shielded sites. The unshielded sites refer to exposure stations directly open to air and seawater actions. These sites are set up mainly to reproduce the real marine environmental actions especially the extreme ones such as the storms and typhoons. For these sites, one needs accurate hydrology data, such as waves and tides, to define the different exposure zones. The shielded sites, on the contrary, are protected, by buildings and facilities, from the strong and extreme environmental actions to increase the safety for both exposed specimens and personnel in operation. For these sites, one needs to clearly specify the different exposure zones created by the shielded conditions. Figure 3.3 illustrate one shielded marine atmospheric site situated in Gulf of Mexico (Port Progresso, Mexico) and another unshielded exposure station on the west artificial island of HZM project (Ling'ding ocean, China).

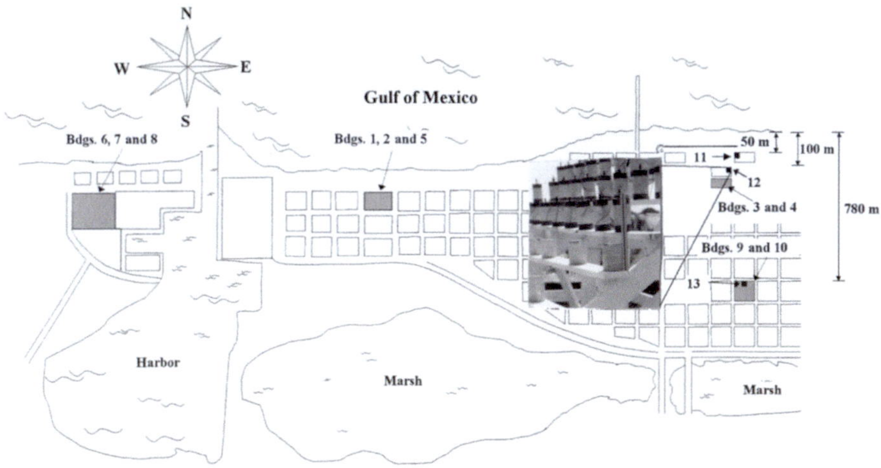


**Fig. 3.1** Geographical location of a typical marine exposure site situated on the coastal area of Gulf of Mexico on the Peninsula of Yucatan, Mexico, and positioning of the tested specimens with respect to the coastal line. *Source* Castro-Borges et al. [13], Fig. 1. Reproduced with the permission of Elsevier

**Fig. 3.2** Seawater salinity survey in Ling'ding ocean, China (left) and meteorological observation equipment on the HZM project exposure site, China (right). *Source* Courtesy of Mr. Jianbo Xiong

**Fig. 3.3** Marine atmospheric exposure site in Mexican gulf, Port Progresso, Mexico (left) and unshielded exposure station for HZM project in Ling'ding Ocean, China (right). Left figure: courtesy of Dr. Pedro Castro-Borges, Right figure source: [24], Fig. 2. Reproduced with permission of Springer

## 3.2 Long-Term Exposure: Factor and Performance

### *3.2.1 Summary on Deterioration Processes*

The central information of long-term exposure pertains to the behaviors of concrete specimens under environmental conditions, and is closely related to the fundamental mechanisms of durability processes from the marine environmental conditions. The different mechanisms have been clarified in the foregoing Sect. 2.1 and Fig. 3.4 summarizes the possible deterioration processes occurring to concrete surface in different exposure zones.

In summary, the deterioration processes are generally divided into surface processes and the inner processes in bulk concrete for different exposure zones. The inner processes are all dominated by the chloride transport via diffusion and binding while the surface processes are closely related to the different exposure zones. In ATM, the surface is subject to the deposition of aerosol chlorides and the carbonation. In SPL, the surface is subject to strong moisture transport (convection) and chloride accumulation by cycles of drying-wetting. In TID, the surface can have leaching and abrasion/erosion in contact with seawater, and also frost actions in cold regions. In SUB, the same leaching and abrasion/erosion processes can be active. Besides, biological growth can be active process for both TID and SUB, cf. Sect. 2. 1.

### *3.2.2 Summary on Influential Factors*

The performance of concrete in marine exposure is affected by several group of factors related to the environmental conditions, the materials and specimens and the service states. The environmental factors refer to the actual impact of seawater or

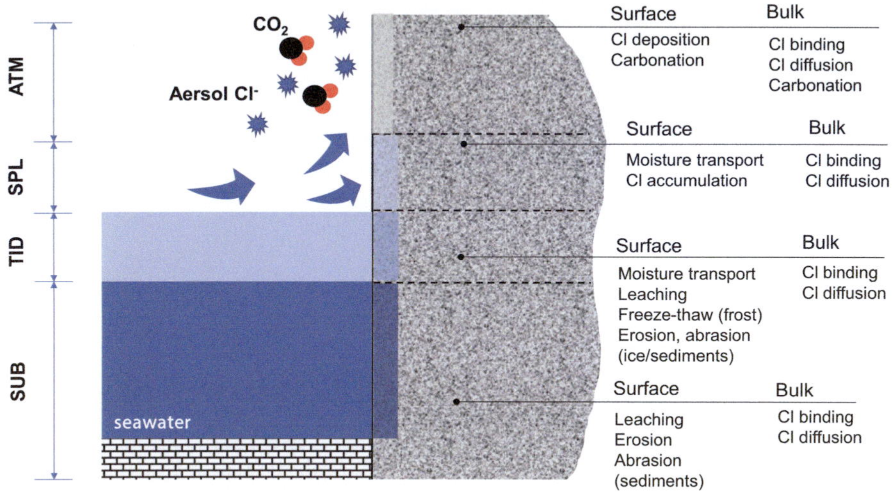


**Fig. 3.4** Summary of deterioration of durability processes of concrete in marine exposure in different zones from ATM, SPL, TID to SUB. *Source* Li et al. [24], Fig. 1. Reproduced with permission of Springer

atmosphere on the surface of concrete specimens, including the effect of specimen orientations and dominating wind directions. The factors of materials and specimens refer to the raw materials, mixture proportioning, and the surface protection materials. The service state factors refer to the loading level and the cracking state of specimens. The relevant mechanisms and the literature results have been reviewed in depth in Sects. 2.2–2.4. It is attempted here to identify the most significant factors, which are to be taken as the essential target factors for data collecting and monitoring during the exposure practice.

The environmental factors include the temperature and salinity of seawater, the contact condition (exposure zone) with seawater, distance from coastline, wind speed and frost intensity. From the literature results, a criticality analysis is given in Table 3.1 for these factors. It can be seen that, except the salinity of seawater, all environmental factors have significant impact on durability processes in marine environments.

The factors of materials and specimens cover the binders, the water to binder ratio (w/b), aggregates and protection of specimens, and the service state factors refer to the loading level and the cracking. Table 3.2 gives a criticality analysis on these factors on the basis of the available literature results. It can be seen that the binder type, together with the water to cement ratio, are significant factors for long-term durability. Furthermore, protection such as silane impregnation treatment on specimens is also of significance. Loading levels and cracking, within the extents specified by technical standards, do not have significant impact on the long-term durability of concrete in marine environments.

**Table 3.1** Criticality analysis for environmental factors for durability performance in marine environments

| Factor | Observation | | Mechanism clarified | Impact on durability | |
|---|---|---|---|---|---|
| | Laboratory | Exposure | | Parameters | Significance |
| Temperature of seawater | Yes | Yes | Yes | Surface concentration, chloride diffusivity | High |
| Salinity of seawater | Yes | Yes | Yes | Surface concentration, chloride diffusivity | Medium |
| Seawater contact (exposure zone) | / | Yes | Partial | Apparent chloride diffusion coefficient | High |
| Distance from coastline | / | Yes | Partial | Deposition rate of chlorides | High |
| Wind speed | / | Yes | Partial | Deposition rate of chlorides | High |

**Table 3.2** Criticality analysis for material and specimen factors for durability performance in marine environments

| Factor | Observation | | Mechanism clarified | Impact on durability | |
|---|---|---|---|---|---|
| | Laboratory | Exposure | | Parameters | Significance |
| Binders | Yes | Yes | Partial | Chloride diffusivity<br>Pore chemistry<br>Chloride binding<br>Microstructure<br>Corrosion risk | High, coupled with w/b |
| Water to binder ratio (w/b) | Yes | Yes | Yes | Chloride diffusivity | High |
| Aggregates | / | Yes | Partial | Chloride ingress | Low |
| Protection of specimens | Yes | Yes | Yes | Chloride ingress | High |
| Loading level, compressive within 0.7 fc | Yes | / | Partial | Permeability | Low |
| Cracking extent, opening within 0.3 mm | / | Yes | Partial | Chloride ingress<br>Steel corrosion extent | Low |

### 3.2.3 Summary on Performance Factors

The impact of all, the environment, the materials and the specimens, as well as service state factors will be manifested through the alteration of the long-term performance. The alteration of such long-term performance has been be characterized by such factors as the mechanical properties, the microstructure and composition, the transport properties, self-healing of cracks, and corrosion resistance of embedded steel as well. Table 3.3 shows the correlation between various factors (environmental, materials and specimens and service state) and the performance (consequence) factors.

**Table 3.3** Relevance of different factors to the deterioration of concrete durability in marine environments (n.r. stands for "not relevant", ↑ for increasing and ↓ for decreasing)

| Factors | | Consequence | | | | |
|---|---|---|---|---|---|---|
| | | Mechanical resistance | Transport resistance | Carbonation resistance | Frost resistance | Corrosion risk |
| Environmental | Salinity (Cl) (seawater) | n.r. | n.r. | n.r. | n.r. | ↑ |
| | Temperature (seawater) | n.r. | ↓ | n.r. | n.r. | ↑ |
| | Drying-wetting (SPL/TID) | ↓ | ↓ | ↓ | ↓ | ↑ |
| | Freeze–thaw (seawater) | ↓ | ↓ | n.r. | ↓ | ↑ |
| Material | Binder type (using SCM) | n.r. | ↑ | ↓ | n.r. | ↓ |
| | w/b ratio (smaller value) | ↑ | ↑ | ↑ | ↑ | ↓ |
| | Pore structure (compact) | ↑ | ↑ | ↑ | ↑ | ↓ |
| | Surface quality (good) | ↑ | ↑ | ↑ | ↑ | ↓ |
| Service | Loading level (compressive) | n.r. | n.r. | n.r. | n.r. | n.r. |
| | Cracking (within 0.3 mm) | n.r. | ↓ | ↓ | ↓ | n.r. |

*Source* adapted from [24], Table 1. Reproduced with permission of Springer

## 3.3 Dispersion and Statistics of Long-Term Exposure Data

An important feature of exposure data is the inherent dispersion. Basically, the dispersion can come from two sources: the inherent dispersion from the randomness related to the environmental actions and the material composition and microstructure, and the artificial dispersion related to the sampling methods, testing methods and the interpretation models for the measured data. This section is dedicated to the inherent dispersion of long-term exposure data, and the basic concepts and tools from engineering statistics are mobilized to describe and handle the dispersion.

### *3.3.1 Probabilistic Description of Long-Term Exposure Data*

The long-term exposure data, as the outcome of durability tests, are associated with significant dispersion, and can be identified through the values of random variables. The random variables can be discrete or continuous. For example, a discrete random variable can be defined for the outcome of steel corrosion inspection: it is assigned a value of "0" if the steel remains passivated, and a value of "1" if the steel is corroded; a continuous random variable can be used to describe the measurements of chloride content in concrete, taking any value within a given interval.

For a discrete random variable, X, which takes on only a finite or countably infinite number of discrete values, the probability mass function is defined as,

$$p_X(x_i) = P(X = x_i), i = 1, \ldots, n \tag{3.1}$$

in which $P$ denotes the occurring probability of $X = x_i$, $n$ can be finite or infinite. The probability distribution function describes the probability $X \leq x$, and relates to the probability mass function through

$$P_X(x) = P(X \leq x) = \sum_{x_i \leq x} p_X(x_i) \tag{3.2}$$

For a continuous random variable, instead of the probability mass function in Eq. (3.1) the probability distribution function is defined as,

$$F_X(x) = P(X \leq x), x \in R \tag{3.3}$$

And the derivative probability function of $F_X(x)$, called the probability density function, is defined by

$$f_X(x) = \frac{dF_X(x)}{dx} \tag{3.4}$$

The probabilistic characteristics are described by the probability distribution function, $P_X(x)$ for discrete variables and $F_X(x)$ for continuous variables. When the form of $F_X(x)$ or $P_X(x)$ is not known in all details, which is the usual case for variables related to exposure practice, it becomes useful to have an estimate of the most important features of the random variables in question. To this purpose, the expected value and variance are often calculated by statistics. The expected value of a random variable $X$, is defined as,

$$E[X] = \sum_i x_i \cdot p_X(x_i) \text{ (discrete) and } E[X] = \int_{-\infty}^{\infty} x \cdot f_X(x)dx \text{ (continuous)} \quad (3.5)$$

The expected value is also called the ensemble average, mean or the first moment of $X$, denoted often by the symbol $\mu_X$. To describe how much a random variable varies, the variance, Var[X] is defined as,

$$Var[X] = \sum_i (x_i - \mu_X)^2 \cdot p_X(x_i) \text{ (discrete) and}$$

$$Var[X] = \int_{-\infty}^{\infty} (x - \mu_X)^2 \cdot f_X(x)dx \text{ (continuous)} \quad (3.6)$$

The dispersion of a random variable can also be described by the standard deviation, $\sigma_X$, with

$$\sigma_X = \sqrt{Var[X]} \quad (3.7)$$

Next, some of most widely used probability distributions are introduced. The most used distribution is the normal distribution, also called Gaussian distribution. Its probability density function writes,

$$f_X(x) = \frac{1}{\sqrt{2}\sigma} \exp\left(-0.5\left(\frac{x-\mu}{\sigma}\right)^2\right) \quad (3.8)$$

where $\mu$ and $\sigma$ are the expected value ($\mu_X$) and the standard deviation ($\sigma_X$) of X, respectively. The probability distribution function corresponding to Eq. (3.8) is given by

$$F_X(x) = \int_{-\infty}^{x} \frac{1}{\sqrt{2\pi}\sigma} \exp\left(-0.5\left(\frac{t-\mu}{\sigma}\right)^2\right) dt \quad (3.9)$$

This integral cannot be evaluated on a closed form. Rather, it is transformed as a standard normal distribution, and calculated by

$$F_X(x) = \int_{-\infty}^{\frac{x-\mu}{\sigma}} \frac{1}{\sqrt{2\pi}} \exp(-0.5t^2)dt = \Phi_X\left(\frac{x-\mu}{\sigma}\right) \tag{3.10}$$

where $\Phi_X$ is the standard normal distribution function. The normal distribution is often used to describe the randomness of concrete cover thickness and aging exponent of chloride diffusion coefficient [25].

If $X = \ln Y$ and $X$ is normally distributed, then the random variable $Y$ follows a lognormal distribution with the probability density function,

$$f_Y(y) = \frac{1}{\sqrt{2}\xi y} \exp\left(-0.5\left(\frac{\ln(y)-\lambda}{\xi}\right)^2\right) \tag{3.11}$$

where $\lambda$ and $\xi$ are the logarithmic mean and logarithmic standard deviation of Y, respectively, which can be obtained from its mean ($\mu_Y$) and standard deviation ($\sigma_Y$) by

$$\xi = \sqrt{\ln\left[1 + \frac{\sigma_Y{}^2}{\mu_Y{}^2}\right]} \text{ and } \lambda = \ln[\mu_Y] - \frac{1}{2}\xi^2 \tag{3.12}$$

and the probability distribution function is given by

$$F_Y(y) = \Phi_Y\left(\frac{\ln(y)-\lambda}{\xi}\right) \tag{3.13}$$

Due to the non-negative property of the lognormal distribution, it is often adopted to describe the randomness of such variables related to exposure practice as the surface chloride concentration, the diffusion coefficient, the strength and the elasticity modulus of concretes [25].

For a random variable, if the knowledge about its distribution is only the upper and lower bounds, it can be described by a uniform distribution with the probability density function and probability distribution function as,

$$f_X(x) = \frac{1}{b-a}, \; F_X(x) = \frac{x-a}{b-a} \tag{3.14}$$

where $a$ is the lower bound and $b$ the upper bound. And its mean and variance are,

$$E(X) = \frac{a+b}{2}; \; Var(X) = \frac{(b-a)^2}{12} \tag{3.15}$$

In the exposure practice, some mineral or chemical contents in concrete raw materials, such as the initial chloride content in binder or concrete, may follow the uniform distribution because the upper or lower bounds are often the first information

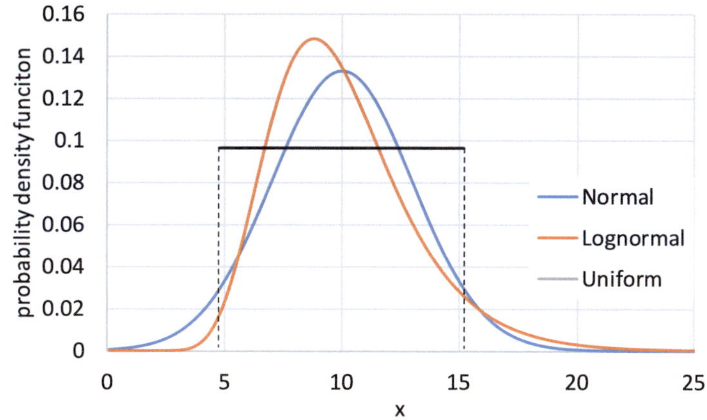


**Fig. 3.5** Probability density functions (Normal, Lognormal and Uniform) with the same mean and variance values ($\mu_X = 10$, $\sigma_X = 3$)

available for these contents. Figure 3.5 shows the three main probability functions with the same mean value and standard deviation.

In durability analysis, it is also important to describe the distribution of the maximum or the minimum value of many samples taken from the parent distribution. They are called extreme distributions and are of three main types, type-I, type-II and type-III. If the upper tail of the parent distribution falls off in an exponential manner, the distribution function, $F_Y$, of the largest value $Y$ of a large number of samples selected at random from the parent population, will be of the form,

$$F_Y(y) = \exp\left[-e^{-a_Y(y-\mathrm{u})}\right] \tag{3.16}$$

in which the parameters $u$ and $a_Y$ are respectively measures of location and dispersion, and the mean and variance of the type-I maximum distribution are related to the parameters $u$ and $a_Y$ as follows:

$$E(Y) = u + \frac{0.5772}{a_Y};\ Var(Y) = \frac{\pi^2}{6a_Y^2} \tag{3.17}$$

The probability density function of the type-I maximum distribution is then:

$$f_Y(y) = a_Y \exp\left[-a_Y(y-u) - e^{-a_Y(y-u)}\right] \tag{3.18}$$

If $Y = \ln Z$ and Y is type-I maximum distributed, then the random variable Z is type-II maximum distributed with the probability distribution function

$$F_Z(z) = \exp\left[-\left(\frac{u}{z}\right)^k\right], z \geq 0, u > 0, k > 0 \tag{3.19}$$

in which the parameters $u$ and $k$ are related to the mean and variance by

$$E(Z) = u\Gamma\left(1 - \frac{1}{k}\right) \text{ with } k > 1 \text{ and}$$
$$Var(Z) = u^2\left(\Gamma\left(1 - \frac{2}{k}\right) - \Gamma^2\left(1 - \frac{1}{k}\right)\right) \text{ with } k > 2 \tag{3.20}$$

where $\Gamma$ is the gamma function defined by

$$\Gamma(k) = \int_0^\infty e^{-u}u^{k-1}du \tag{3.21}$$

The type-I and type-II distributions are usually used to describe the maximum of many similar but independent events, e.g. the maximum daily wave height at a particular site. Type-III is considered to describe the smallest extreme, it arises when the parent distribution is limited to the left at a value $\epsilon$. The probability distribution of the minimum Y of n independent and identically distributed random variables approaches the form.

$$F_Y(y) = 1 - \exp\left[-\left(\frac{y-\epsilon}{u-\epsilon}\right)^k\right] \text{with } y \geq \epsilon, k > 0, u > \epsilon > 0 \text{ as n} \to \infty \tag{3.22}$$

The mean and variance of Y are

$$E(Y) = \epsilon + (u - \epsilon)\Gamma\left(1 + \frac{1}{k}\right) \text{ and}$$
$$Var(Z) = (u - \epsilon)^2\left(\Gamma\left(1 + \frac{2}{k}\right) - \Gamma^2\left(1 + \frac{1}{k}\right)\right) \tag{3.23}$$

### 3.3.2 Statistical Estimation of Distribution Parameters

The probabilistic models introduced previously can be used for describing long-term exposure data. These probabilistic models are determined through distribution parameters, e.g., $u$ and $\sigma$ for normal distribution, $\lambda$ and $\xi$ for lognormal distribution, and $u$, $\epsilon$ and $k$ for extreme type-III distribution. In practice, however, these distribution model parameters are unknown since the probability distribution that characterizes measurements from the population is unknown. In that case, the statistics of measurements are firstly obtained, and then the distribution parameters are determined through the relationship between them and the statistics of samples by assuming the samples follow a specific distribution.

Sample mean and sample variance are two main point estimates. If $X_1, \ldots, X_n$ is a sample of measurements from a probability distribution with a mean $\mu_X$ and a variance $\sigma_X{}^2$, then the sample mean

$$\overline{X} = \frac{1}{n}\sum_{i=1}^{n} X_i \tag{3.24}$$

is an unbiased point estimate of the actual population mean, i.e.

$$\widehat{\mu}_X = \overline{X} \tag{3.25}$$

and the sample variance

$$S^2 = \frac{1}{n-1}\sum_{i=1}^{n} \left(X_i - \overline{X}\right)^2 \tag{3.26}$$

is an unbiased point estimate of the actual population variance, i.e.

$$\widehat{\sigma}_X{}^2 = S^2 \tag{3.27}$$

Once the sample mean and sample variance are estimated, the distribution parameters of probabilistic models with two parameters can be determined, e.g. normal distribution, lognormal distribution, uniform distribution, extreme type-I and type-II distributions.

The sample mean is also a random variable, and according to the central limit theorem, the sample mean $\overline{X}$ has an approximate normal distribution with the mean of $\overline{X}$ and the standard deviation of $\frac{\widehat{\sigma}_X}{\sqrt{n}}$, i.e.

$$\overline{X} \sim Normal\left(\overline{X}, \frac{\widehat{\sigma}_X}{\sqrt{n}}\right) \tag{3.28}$$

This distribution is exact if the random variables $X_i$ are normally distributed. Equation (3.28) demonstrates that the estimate of population mean becomes more accurate as more samples are obtained. Let us demonstrate this with the following example. Suppose 100 samples of surface chloride content (unit: %binder) are obtained, as shown in Fig. 3.6.

The mean and the standard deviation of these 100 samples are calculated, according to Eqs. (3.24) and (3.26), as $\overline{X} = 2.00$, $S = 0.39$. The probability distribution of these samples is plotted in Fig. 3.7, together with the normal distribution with mean of 2.00 and standard deviation of 0.39. It can be seen that these samples basically follow a normal distribution.

If only the first 5, 10, 20 and 40 samples are available, the sample mean and sample deviation are also obtained, as seen in Table 3.4, and the probability density

| | | | | | | | | | | | | | | | | | | | |
|---|---|---|---|---|---|---|---|---|---|---|---|---|---|---|---|---|---|---|---|
| 1.9 | 1.7 | 1.9 | 1.6 | 2.3 | 2.5 | 1.7 | 1.3 | 1.6 | 1.8 | 1.8 | 2.2 | 1.7 | 1.9 | 2.1 | 1.3 | 2.6 | 2.2 | 1.6 | 2.4 |
| 1.8 | 1.8 | 2.0 | 1.4 | 1.9 | 2.6 | 1.5 | 1.7 | 2.2 | 2.3 | 2.2 | 2.2 | 2.4 | 2.1 | 1.7 | 2.2 | 2.4 | 2.7 | 2.2 | 2.0 |
| 1.8 | 2.1 | 2.2 | 1.5 | 2.2 | 2.0 | 2.1 | 2.7 | 2.0 | 2.4 | 1.8 | 2.7 | 2.2 | 1.9 | 1.6 | 2.2 | 1.9 | 2.4 | 1.1 | 2.2 |
| 1.6 | 2.4 | 2.1 | 2.5 | 1.9 | 2.7 | 1.4 | 1.9 | 1.3 | 2.4 | 2.5 | 2.1 | 2.3 | 1.8 | 1.7 | 2.5 | 2.3 | 2.4 | 3.0 | 1.7 |
| 2.4 | 1.8 | 1.5 | 1.6 | 1.4 | 1.5 | 1.6 | 1.7 | 1.9 | 1.9 | 2.5 | 2.5 | 1.5 | 1.7 | 1.8 | 1.6 | 2.0 | 2.2 | 2.4 | 1.7 |

**Fig. 3.6** Samples of chloride content measurements (% binder) for demonstration

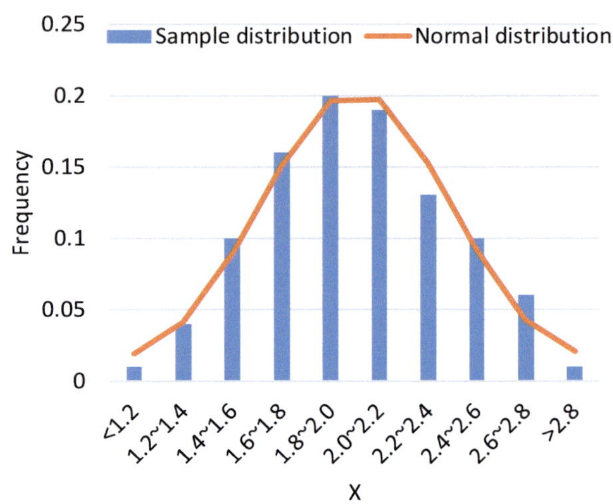


**Fig. 3.7** Probability distribution of samples from all 100 measurements

**Table 3.4** Sample means and sample standard deviations from different samples

| Sample number | 5 | 10 | 20 | 40 | 100 |
|---|---|---|---|---|---|
| Mean | 1.879 | 1.830 | 1.905 | 1.985 | 2.001 |
| Standard deviation | 0.268 | 0.350 | 0.371 | 0.363 | 0.389 |

functions according to the estimated mean and standard deviation are plotted in Fig. 3.8, demonstrating the improved model accuracy with the number of samples available.

## 3.4 Sampling and Testing

The sampling procedure and the relevant test methods are crucial information in view of correct interpretation of exposure data and their comparison and sharing among different exposure sites. This section is dedicated to the sampling procedure and test methods adopted in the exposure practice so far. Since neither the sampling nor the

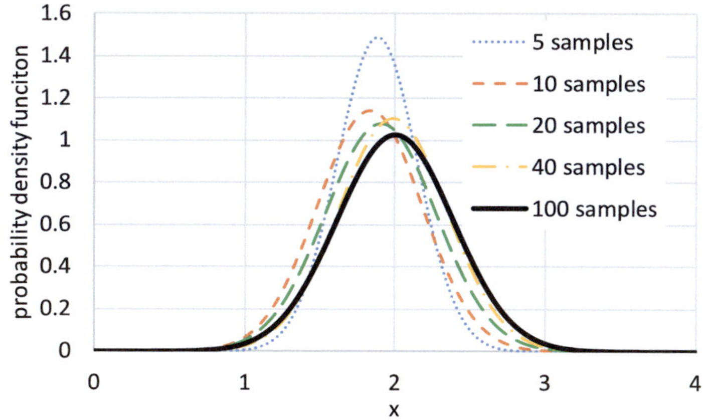


**Fig. 3.8** Estimated probability density functions using the sample mean and sample deviation of different number of samples

test methods are normalized, the best practice is introduced in the following on the basis of the available experiences.

### *3.4.1 Sampling and Conditioning*

Before sampling, the biological growth on the exposed surface of concrete specimens should be removed carefully. Samples of concrete materials are usually taken from the surface to the inner part of specimens, and dry-coring or dry-cutting are used to avoid the dilution of chlorides from sampling operation [15, 22]. One typical chemical composition analysis is to determine the chloride profiling analysis through titration and acid-soluble or water-soluble testing. For this analysis, a certain mass of powder sample should be collected from the specimen at different depths, and sieved to obtain a homogeneous sample of particles.

So far, two methods are used to prepare powder samples for the chemical analysis. The first option is to generate powder samples from mortar in concrete, which involves slicing concrete at different depths, separating the large aggregates, and grinding and sieving the mortar to obtain powder samples. However, on the one hand small aggregates are difficult to remove, and on the other hand chloride ions at the mortar-aggregate interface may be removed.

The second method is to generate the powder samples directly by drilling vertically into concrete specimens and collecting the powder samples. The first method somehow has the difficulty to avoid small coarse aggregates while the second needs a high enough amount of powder to remain representative. Technical standards impose the necessary amount of powder samples with respect to the maximum aggregate size. The standards ASTM C1152 [6] and ASTM C1218 [7] specify that when the

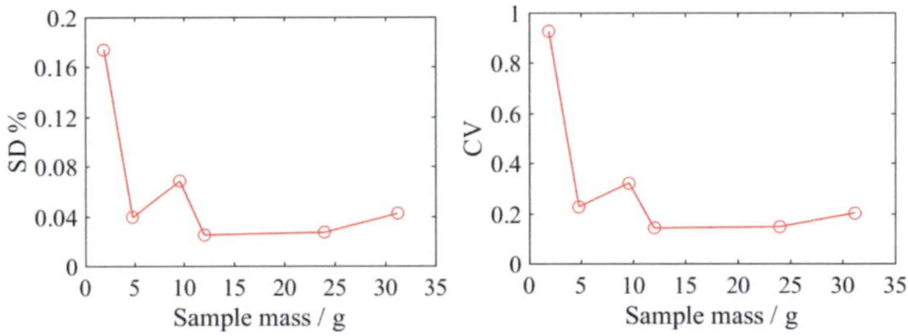


**Fig. 3.9** Influence of the sample mass on chloride content analysis result: standard deviation (left) and coefficient of variance (right). Data extracted from ASTM C1152 [6] and ASTM C1218 [7]

coarse aggregate size is larger than 25 mm the sample weight should be at least 20 g. The relevant research results on concrete specimens with maximum aggregate of 25 mm are shown in Fig. 3.9: the standard deviation and coefficient of variation in chloride content decrease with the sample volume, and when the sample weight exceeds 12 g, or volume exceeds 5 $cm^3$, the standard deviation and coefficient of variation converge.

For powder samples, the particle size should also be considered as an influential factor for analysis results. Normally, the particle size of the powder sample ranges from 0.16 to 0.85 mm for chloride concentration (content) analysis. The standard ASTM C1152 [6] requires that the powder passes through a 0.85 mm sieve, and the RILEM TC 178-TMC [35] recommends a particle size of less than 0.315 mm. Arya et al. [5] found that chloride test results are the same when the powder particle size is no more than 0.60 mm. The relevant study for the standard JGJ/T 322-2013 [17] on mortar powder with particle sizes of less than 0.08 mm, 0.16 mm, 0.315 mm, and 0.63 mm showed that the water-soluble chloride content was highest in powders of 0.16 mm with small difference among all sizes, cf. Fig. 3.10.

Some state-of-art practice in exposure sites is summarized here. The Zhanjiang exposure station in China uses rotary impact drill to perform the drilling to collect the powder samples from concrete specimens. The sampling surface size is larger than 3 times of the maximum size of coarse aggregates. At least 10 g powder sample is collected for each layer if the aggregate size smaller than 25 mm and 20 g powder sample for aggregate size larger than 25 mm. The particle size of powder samples is controlled to less than 0.16 mm, and the standard NT Build 443 [29] is followed for the procedure.

The Port of Progreso exposure site, Mexico follows the standard NMX-C-169 [30] for the procedure. Depending on the type of sample, on site or in laboratory, one or more types of samples are extracted, where one end of the specimen corresponds to the surface exposed to the atmosphere while the other end corresponds to the internal part of concrete. Once concrete cores are obtained, the cores are marked every 10 mm from the air-exposed end and transverse slices are cut out along the specimen. Each slice is divided into smaller pieces with an appropriate mill, or with an impact tool.

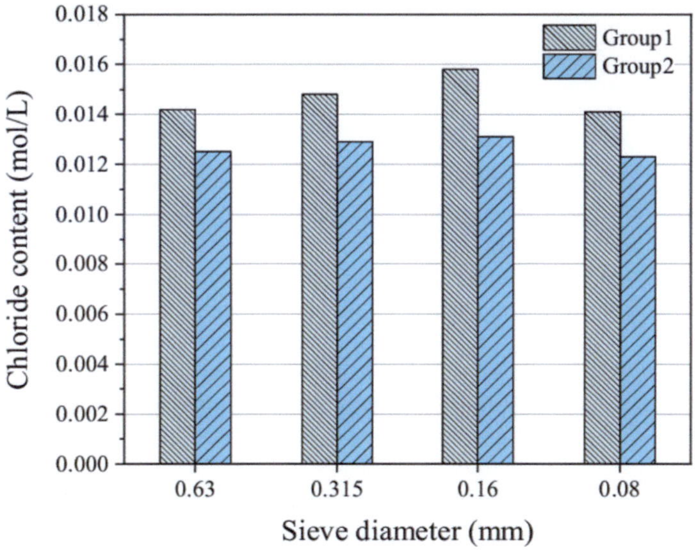


**Fig. 3.10** Water-soluble chloride content in two groups of concrete tested on mortar powder sample with different particle sizes. Data extracted from JGJ/T 322-2013 [17]

Subsequently, the sample is ground to powder passing through a 297 μm (No. 50) sieve. At least 10 g of powder sample is collected for each layer.

### *3.4.2 Testing*

So far, no general protocol for testing on the exposure specimens has been established. The main testing for durability performance consists of carbonation depth and chloride ingress. The carbonation depth is usually determined by the spray of phenolphthalein solution [31] or other reagents [14], and the chloride ingress is usually quantified through profiling method which determines the chloride content in powder samples collected from different depths of specimen. Here the best practice for profiling method is reviewed. The chloride content or concentration can be determined either through water-soluble method [6, 10, 33] or acid-soluble method [1, 7, 11, 34]. The potentiometric titration is usually used to measure the chloride concentration of the liquid solution filtered from the dispersed powder sample.

The water-soluble method extracts chlorides from samples with the following procedure: soak the sample in distilled water, filter it through filter paper, and then add an acidic solution (nitric acid) to avoid precipitation. Then the sample is ready for the subsequent potentiometric titration. The main differences among different extraction methods on operation levels are the soaking time, whether shaking or boiling is used and how sulfide interference is handled. The standard ASTM C1218

[6] involves boiling the soaking sample for 5 min and then letting it stand for 24 h. After comparing several different water extraction methods, the RILEM TC 178-TMC [33] recommended a simple soaking time of 3 min of samples on a magnetic agitating plate. In Zhanjiang exposure site (China), the soaked samples are shaken for 1–2 min and then left standing for 24 h. Muralidharan et al. [28] showed that the standing time 24–48 h had no notable effect, the mixing with boiled/boiling water released the most chloride contents, and the shaking 1 h method gave the lowest measurements. However, the boiling treatment during water extract is problematic because such treatment distorts the conventional dissolution equilibrium between the aqueous chlorides and chlorides in solid phases in concrete pore environment.

The soaking of powder samples from mortar or concrete may trigger the leaching of bound chlorides in solid phases. Chaussadent and Arliguie [16] found that when the soaking time exceeds 3 min the calcium chloroaluminate ($C_3A{\cdot}CaCl_2{\cdot}10H_2O$) begins to dissolve, releasing bound chlorides from the solid phases. To prevent this, the soaking time for the water-soluble test is fixed at 3 min. Besides, sulfides may influence the measurement of chlorides in powder samples, and ASTM C1218 [6] and RILEM TC 178-TMC [35] use hydrogen peroxide for treatment.

The acid-soluble method for extracting chlorides from powder samples is similar to water-soluble method except that the powder sample is soaked entirely in nitric acid ($HNO_3$). The main differences at operation level among the extraction methods are the soaking and boiling time. RILEM TC 178-TMC [34] recommended to heat the suspension (samples soaked in nitric acid) 1 min after boiling and another 1 min after adding standard 0.05 M $AgNO_3$ solution. Figure 3.11 shows the measurements with different extraction methods from Zhanjiang exposure site, China [19]. The four extraction methods are, soaking for 24 h, boiling for 5 min and soaking for 24 h, soaking with shaking for 6 h, boiling for 5 min and soaking for 6 h. The measured acid-soluble chloride contents are nearly the same for the four methods, indicating that they are all effective in extracting chlorides from mortar.

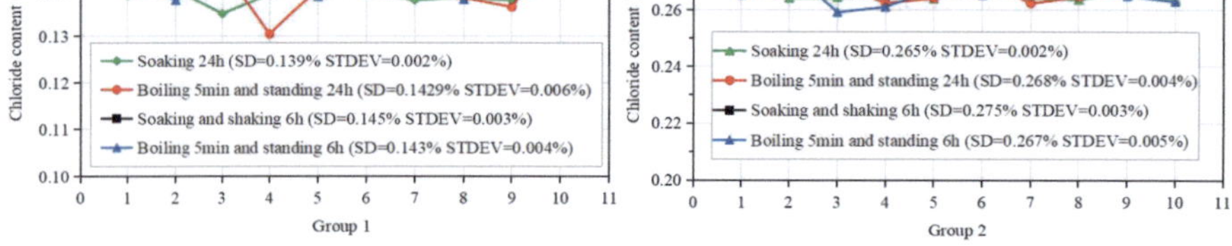


**Fig. 3.11** Chloride content in two groups of mortar powder samples tested by different extraction methods using acid-soluble method from Zhanjiang exposure site, China. Data extracted from technical report [19]

Apart from the chloride profiling method [27, 35] applied μXRF method for measuring chlorides in concrete. The raw data from μXRF are the numbers of fluoresced X-rays within a certain energy window including background radiation. To quantify the chloride concentration, a calibration process is needed to convert the X-ray counts to chloride concentration. A threshold value for chlorides of 200 ppm was adopted to mark the depth of chloride penetration into concrete.

The Zhanjiang exposure station, China, adopts acid-soluble method to determine the chloride content in concrete. For each powder sample, a minimum amount of 10 g is required and the soaking solution is nitric acid (1 + 7) and the soaking time is 24 h. The potentiometric titrator method is used to determine the chloride content following the standard JGJ/T 332 [17]. The port of Progreso exposure site, Mexico, uses also acid-soluble method and an amount of 2.4 g is required for each powder sample. The soaking solution is 0.6 M nitric acid. The soaking tine is 15 min of agitation plus 1 h of sedimentation. The potentiometric titrator method is used following the standard NMX-C-523 [32]. However, as demonstrated by the results in Table 3.4 and Fig. 3.8, the dispersion of testing results from the sampling and testing is yet to be considered and quantified in the future, answering such questions as how many samples are enough to obtain stable and meaningful results for target variables and parameters. Actually, this belongs to the interdisciplinary domain between the probability theory and exposure practice, which opens future directions for research.

### *3.4.3 Corrosion of Rebar*

Monitoring and detecting the electrochemical state of embedded rebars are highly related to the long-term exposure practice but very different from the testing of concrete/mortar samples. Thus, the testing of the corrosion of rebars is reviewed separately here. The standard electrochemical methods have been developed for testing the corrosion of embedded rebars in concrete. RILEM TC 154-EMC has developed test methods for electrical resistivity [2], half-cell potential [4] and corrosion rate measurement [20]. The most widely used method would be the half-cell potential measurement [4, 8]. The concept behind is as follows: When an oxidation reaction occurs on the surface of the steel bar, an electron flow is generated and a potential difference is formed between the surface of the rebar and the surrounding environment. By measuring the value of this potential difference, the degree of corrosion of the steel bar can be determined. The Cu-$CuSO_4$ electrode and the saturated calomel electrode are the most used electrodes in practice. Further, it was pointed out that the threshold values recommended for the corrosion states should be treated with precaution because the moisture state can have notable impact on these potential values, e.g. a wetting event can push the whole potential field to more negative values by 100 mV [4].

The electrochemical state of steel can be detected by multiple embedded probes and sensors [40] while portable devices have also been applied to detect the rebar corrosion in concrete. The hand-held non-destructive device RapiCor was applied by

[38]. Similar to the typical galvanostatic pulse measurement, the instrument firstly measures the corrosion potential $E_{corr}$ by the reference electrode placed at the center of the electrodes unit and afterwards imposes galvanostatic currents $I_{CE}$ and $I_{GE}$ through the counter electrodes "CE" and the guard electrodes "GE" to the steel bar embedded in concrete. Immediately after having imposed the currents $I_{CE}$ and $I_{GE}$, the data acquisition system records the signal responses of potential $\Delta E_a(t)$ at a time interval of less than 0.02 s. The recorded potential-time curve is directly available to calculate such parameters as ohmic resistance $R_\Omega$, polarization resistance $R_p$, etc.

## 3.5 Presentation of Marine Field Exposure Data

On the basis of the above analysis of all information concerning long-term exposure sites and exposure practice, this section is dedicated to the normalization of data presentation for exposure practice. This normalization has two-folded purposes: complete the essential information on exposure practice to support better the interpretation of exposure data, and to facilitate the data comparison and exchange for the sake of valorizing the long-term exposure practice.

### *3.5.1 Summary Review of Information Relevant to Marine Exposure Practice*

During the TC works, the experts were mobilized to collect the relevant information on exposure sites and exposure practice. For exposure sites, the collected information includes the climate, siting geology and environments, support facilities for exposure practice, the fabrication of specimens, the position and placement of specimens and the sample collecting schemes. In total, nine exposure sites (stations) were collected and given in Table 3.5. For exposure practice, the information collected includes the test methods and measured indexes for concrete, rebars, protection and others, and the data recording, storage and implementation. In total eight exposure sites were collected and detailed in Table 3.6. The information presented in these two tables constitute the first basis for the further normalization of exposure data in the following.

### *3.5.2 Essential Information for Presentation of Marine Exposure Data*

From the collected information and the state-of-the-art practice, it is attempted to establish the minimum ensemble of information describing a marine exposure for

**Table 3.5** Information collected on the main technical characteristics for marine exposure sites

| Exposure site | Climate | Siting conditions | Support structure | Specimen | | |
|---|---|---|---|---|---|---|
| | | | | Fabrication | Positioning and protection | Sample collecting |
| Zhanjiang Port (China) [41] | Subtropical marine, southeastern coast of China | Marine coast, Shielded conditions | Wharf structure, RC frame | 150 mm × 150 mm × 300 mm; 28 d laboratory curing; labelling after zone-concrete-purpose | Specimens placed in different exposure conditions relative to seawater, positioned and protected by stainless steel cages; facilities designed to withstand typhoons and ship collisions | Samples collected at pre-determined intervals; numbers of samples collected according to specimen size and purpose: corrosion of rebars, chloride ingress, ageing of coatings |
| Hong Kong-Zhuhai-Macau Bridge (China) [23] | Subtropical marine, southeastern coast of China | Offshore, unshielded conditions | Independent RC frame, built for exposure | 150 mm × 150 mm × 150 mm (plain concrete), 150 mm × 150 mm × 300 mm (RC specimen); 28 d laboratory curing; labelling after zone-concrete-purpose | Same as above | Same as above |
| Träslövsläge (Sweden) [36] | Nordic marine, west coast of Sweden | Marine coast, shielded by breakwater in harbor | Floating pontoons | 1000 mm × 700 mm × 100 mm; 14 d laboratory curing; labelling with stainless steel plate | Specimens mounted on the floating pontoon with the half of the length submerged in the sea | Chloride profiles after exposure for 0.5–1a/2a/5a/10a/20a, and corrosion conditions after exposure for 13a/20a [9] |

(continued)

**Table 3.5** (continued)

| Exposure site | Climate | Siting conditions | Support structure | Specimen | | |
|---|---|---|---|---|---|---|
| | | | | Fabrication | Positioning and protection | Sample collecting |
| Fehmarn Belt (Denmark) [37] | Nordic marine, south coast of Denmark | Marine coast, shielded by breakwater in harbor | Steel rigs fixed on sheet pile wall | 2000 mm × 1000 mm × 200 mm; 43–49 days curing; labelling with imprint on top and marking on stainless steel eyebolts for lifting | Specimens mounted in steel rigs with the 700 mm above and 1300 mm below normal water level | Chloride and calcium profiles for 15 mix designs after exposure for 0.5a/2a/5a/10a, and continuous monitoring of corrosion for 3 mix designs |
| Cape Peninsula (South Africa) [26] | Temperate, eastern (Granger bay) and western (Simonstown) peninsula | Marine coast: shielded by a dockyard (Simonstown), unshielded with strong wind (Granger bay) | Sea wall | Plain concrete block specimens; different curing methods;<br>28 days (age at the start of exposure) | Secured through chains; specimens exposed to upper tidal and spray zones, oriented towards the worst combination of drying and seawater spray | Chloride content and durability indicators, on predetermined time basis |

(continued)

**Table 3.5** (continued)

| Exposure site | Climate | Siting conditions | Support structure | Specimen | | |
|---|---|---|---|---|---|---|
| | | | | Fabrication | Positioning and protection | Sample collecting |
| Wheat Island, Qingdao (China) | Temperate monsoon climate, northeast coast of China | Marine coast, Shielded conditions | Natural reefs and RC frame | 100 mm × 100 mm × 100 mm (plain concrete); 100 mm × 100 mm × 300/400 mm (RC specimen); 28d laboratory curing; labelling after zone-concrete-purpose | Same as Zhanjiang Port | Same as Zhanjiang Port |
| Progreso Port, Yucatán, (México) [12] | Tropical climate, coast of Mexico Gulf | 50–780 m from the coast, total exposure to marine air | Aluminum frames | 150 mm × 150 mm × 300 mm (plain concrete/RC); 28d laboratory curing; labelling after number-w/c ratio-prevailing or Non Prevailing wind | Specimens with one 150 mm × 300 mm surface oriented toward the prevailing winds | Carbonation depth, profiles of chlorides and sulfates measured on predetermined basis; half-cell potential and corrosion rates measured monthly using the rebars in RC prisms |

(continued)

**Table 3.5** (continued)

| Exposure site | Climate | Siting conditions | Support structure | Specimen | | |
|---|---|---|---|---|---|---|
| | | | | Fabrication | Positioning and protection | Sample collecting |
| NPRA (Norway) | Nordic marine, Norwegian coast | Marine coast, shielded by breakwater in harbor (Solsvik, Austefjorden) | Quay | 300 mm × 300 mm × 3000 mm (Sandnessjøen, Solsvik), 190 mm × 390 mm × 2700 mm (Austefjorden); minimum 14d curing; labelling according to composition | Specimens placed partially submerged in seawater with parts in air, tidal zone, and submerged; secured through chains (Sandnesssjøen, Solsvik) or steel frame (Austefjorden) | Core extraction and determination of chloride content (profiles) at predetermined intervals; Regular measurements (Sandnessjøen) or instrumental monitoring (logging) of signals (Solsvik, Austefjorden) for corrosion activity above, in and below the tidal zone |
| Huelva (Spain) [3] | Mediterranean climate, Southwest coast of Spain | Marine coast, shielded beach of 200 m long, closed to public | No | Reinforced concrete blocks of 50 × 50 × 2000 cm made with different cement types and reinforcements | Blocks half-buried in sands (1 m), half-exposed to tidal actions (1 m) | Cores drilled and chloride profiles determined by grinding powder; Corrosion rate and resistivity measured since some years ago |

*Source* Li et al. [24], Table 3. Reproduced with permission of Springer

**Table 3.6** Data-relevant information collected on the marine exposure sites

| Exposure site | Test methods and measured indexes | | | | Data management | |
|---|---|---|---|---|---|---|
| | Concrete | Rebar | Coating/silane | Others | Recording and storage | Implementation and use |
| Zhanjiang Port (China) [41] | Chloride profiling from ground powder (ASTM C1152); measured indexes: chloride content, apparent chloride diffusion coefficient, surface chloride concentration | Corrosion rate of rebars in concrete (ASTM C876); measured indexes: corrosion potential, corrosion current, polarization resistance | Bonding force coating/ concrete (ASTM D7234), depth of silane and water absorption of concrete (ASTM C1585); measured indexes: bonding force, impregnation depth, water absorption | Resistance to chloride penetration in concrete mixed with hydrophobic pore-blocking agent (ASTM C1585/C1152); measured indexes: water absorption, apparent chloride diffusion coefficient | Manual recording and file storage on paper formats (before 1990s), and computerized recording and storage in specific database developed (after 1990s) | Data centralized Storage and implementation by CCCC; data accessible to authorized parties; support to update relevant technical specifications |

(continued)

**Table 3.6** (continued)

| Exposure site | Test methods and measured indexes | | | | Data management | |
|---|---|---|---|---|---|---|
| | Concrete | Rebar | Coating/silane | Others | Recording and storage | Implementation and use |
| HongKong-Zhuhai-Macau Bridge (China) [23] | Same as above | Same as above | Same as above | Same as above | Computerized recording and storage in specific database; included into integral HZM project mega-database of service state control | Data storage and implementation by CCCC (owned by HZMBA); data accessible to authorized research units; support to HZM project |

(continued)

**Table 3.6** (continued)

| Exposure site | Test methods and measured indexes | | | | Data management | |
|---|---|---|---|---|---|---|
| | Concrete | Rebar | Coating/silane | Others | Recording and storage | Implementation and use |
| Träslövsläge (Sweden) [36] | Chloride profiling from ground powder (NT BUILD 443); measured Indexes: chloride content, apparent chloride diffusion coefficient, surface chloride content | Corrosion condition of rebars in concrete (ASTM C876; RapiCor); measured indexes: corrosion potential, corrosion rate, resistivity of cover concrete | Not included | Not included | Manual recording and file storage | Data published in technical reports; data accessible to public in form of electronic tables |

(continued)

**Table 3.6** (continued)

| Exposure site | Test methods and measured indexes | | | | Data management | |
|---|---|---|---|---|---|---|
| | Concrete | Rebar | Coating/silane | Others | Recording and storage | Implementation and use |
| Fehmarn Belt (Denmark) [37] | Chloride and calcium profiling from ground powder (NT BUILD 443); measured indexes: chloride and calcium content, apparent chloride diffusion coefficient, surface chloride content, compressive strength, microstructure by petrography | Corrosion condition of rebars in concrete (3 mixes) (ASTM C876); measured indexes: corrosion potential, corrosion rate | Not included | For three mixes: corrosion initiation on anode ladders. Self-healing of cracks and corrosion initiation of intersecting rebars | Automatic monitoring and file storage of corrosion data. Manual recording and file storage of concrete exposure data | Data published in technical reports; data accessible to public in form of electronic tables. Data available at: www.concreteexpertcentre.dk |

(continued)

**Table 3.6** (continued)

| Exposure site | Test methods and measured indexes | | | | Data management | |
|---|---|---|---|---|---|---|
| | Concrete | Rebar | Coating/silane | Others | Recording and storage | Implementation and use |
| Wheat Island, Qingdao (China) | Chloride profiling from ground powder (JTS/T 236-2019); measured indexes: same as Zhanjiang Port | Same as Zhangjiang Port | Depth of silane and water absorption of concrete (ASTM C1585); measured indexes: impregnation depth, water absorption | Same as Zhangjiang Port | Recording and storage in specific database developed (built in 2010) | Data storage and implementation by QUT; data accessible to authorized units; support to project (Qingdao Bay Bridge) |

(continued)

**Table 3.6** (continued)

| Exposure site | Test methods and measured indexes | | | | Data management | |
|---|---|---|---|---|---|---|
| | Concrete | Rebar | Coating/silane | Others | Recording and storage | Implementation and use |
| Port Progreso, Yucatán, (México) [12] | Carbonation front (NMX-C-515); profiles of water-soluble chlorides (RILEM TC 178-TMC); Resistivity (NMX-C-514); measured indexes: carbonation depth, chloride diffusion coefficient, resistivity | Corrosion conditions for rebars in concrete (NMX-C-501, NMX-C-495); measured indexes: corrosion potential, corrosion current, polarization resistance | Not included | Physical properties of concrete [18]; measured indexes: water absorption rate, effective porosity | Manual recording and file storage on paper formats, then computerized recording and storage in specific database developed | Support to Mexican NMX standards and projects in the same region |

(continued)

**Table 3.6** (continued)

| Exposure site | Test methods and measured indexes | | | | Data management | |
|---|---|---|---|---|---|---|
| | Concrete | Rebar | Coating/silane | Others | Recording and storage | Implementation and use |
| NPRA (Norway) | Chloride profiling from ground powder (spectrophotometric or potentiometric method); measured indexes: same as Träslövsläge station | Corrosion conditions of rebars in concrete; measured indexes: corrosion potential, electrical resistance | Not included | Resistance to chloride penetration in (1) pre-cracked beams, (2) concrete mixed with corrosion inhibitor, (3) concrete mixed with hydrophobic agent, (4) concrete mixed with latex (Sandnesssjøen); measured indexes: same as concrete tests | Manual recording and file storage on paper formats, then computerized recording and storage | Data used for revision of NPRA rulebooks; chosen data published in technical reports and/or journals |
| Huelva (Spain) [3] | Chloride and XRD profiling; MEASURED INDEXES: CHLORIDE content, apparent chloride diffusion coefficient, surface chloride content | Corrosion rate through linear polarization; measured indexes: corrosion rate | Not included | Several cement types and concrete mixes with galvanized and stainless steel rebars | Data recorded every year until around 20 years | Data collected by authors and support for general research and engineering projects |

*Source* Li et al. [24], Table 4. Reproduced with permission of Springer

both the completeness of dataset and the convenience of data comparison and sharing. Table 3.7 presents the factors related to the characteristics of exposure sites and the exposure practice, and splits the factors into three groups: the environment, the material and the performance groups. Note that this table provides a minimum collection of information relevant to marine field exposure for conventional engineering use and additional information can be necessary for specific exposure or research purposes.

**Table 3.7** Minimum collection for standard presentation of long-term marine exposure data

| Group | Factors and indexes | |
|---|---|---|
| Environment | Exposure station description | Station owner/operator (ID), exposure location (ID), opening year, status of station, exposure conditions provided, capacity of exposure station |
| | Meteorological and hydrological conditions | Atmospheric temperature, relative humidity, wind speed, precipitation, exposed zones (atmospheric, splash, tidal submerged), seawater salinity, temperature, chloride concentration and sulfate concentration |
| | Additional environmental conditions | Ambient temperature variation, relative humidity variation, additional seawater chemistry, chloride deposition rate |
| Materials | Concrete specimen label | Specimen ID, mix ID, specimen geometry, casting date, exposure date, exposure location/station (ID) |
| | Concrete mix proportioning and properties | Mix proportion, air content and chloride content of the fresh concrete, density and slump of fresh concrete, laboratory production and curing, compressive strength of concrete at 28d |
| | Raw materials and other properties | Type, chemical composition and physical properties of cement and mineral admixtures; type, maximum size and grade curve of aggregates; specimen surface cracking (if apply) |
| Performance | Concrete mechanical properties | Compressive strength; dynamic modulus of elasticity (if apply) |
| | Chloride profiles | Chloride profile, test method of chlorides |
| | Steel profiles (for reinforced concrete) | Corrosion potential, corrosion rate, resistivity tensile strength (if apply) |
| | Additional performance data | Profiling of other ion species, carbonation/leaching depth, microclimate of specimen surface, concrete cover compactness, concrete surface cracking |

*Source* Li et al. [24], Table 2. Reproduced with permission of Springer

**Table 3.8** Basic data structure for exposure sites and exposure data

| Table ID | Data description |
|---|---|
| Exposure SITES_ DATA | **Exposure station number**, name, geographic location, owner, opening time, operating status (closed/active), exposure zones provided, capacity of specimen accommodation |
| Exposure sites_ environment data | **Exposure station number**, annual temperature, annual relative humidity, annual precipitation, annual average wind speed, annual average seawater temperatures, position of exposure zones (ATM, SPL, TID, SUB), salinity of seawater, chloride and sulfate concentrations, other optional conditions (e.g. chloride deposition rate) |
| Concrete specimen_data | **Specimen ID**, mixing ID, specimen geometry, casting date, exposure date, proportioning and properties (cement, aggregates etc.), raw materials (cement type, minerals), specimen state and protection (cracking, coating) |
| Exposing conditions | **Exposure condition number**, specimen_ID, chloride profiling date, exposure station number, exposure zone, surface protection, chloride determination method |
| Chloride profiling | **Exposure condition number**, chloride profile (depth and chloride concentration), regressed diffusion coefficient, regressed surface chloride concentration |

### 3.5.3 *Database for Marine Exposure Data*

Following the above normalized presentation of exposure data in marine environments, a preliminary database was built from the collected data during the TC works. Among these data were included more than 1000 chloride profiles of concrete specimens from the exposure stations in Mexico, Sweden, Spain, South Africa and China. The basic data structure is listed in Table 3.8 through five tables. Currently, a total of 626 profiles have been uploaded into the database, including Cape Town, South Africa (118 profiles), HZM Station, China (48 profiles), Träslövsläge Station, Sweden (235 profiles), and Yucatan Station, Mexico (235 profiles).

Some basic functions were developed for this database, including the search of specimens via the mixture (water to binder ratio, cement content etc.). Once the specimen ID is obtained, it can be used as handle to retrieve the mixture data, the chloride profiles and the exposure site information. For chloride profiles, a simple plot of chloride content in terms of depth can be presented of which the original data can be downloaded. Certainly, these basic functions will be improved during the use of this database, and the data structure can be also optimized in the future.

## 3.6 Chapter Summary

This chapter focuses on the information relevant to the exposure sites (stations) and the exposure practice as well. The information begins with the characteristics of exposure sites, continues with the factors related to the durability performance of exposure specimens, elaborates the statistical nature of the collected data, stresses sampling and testing in data collection, and finishes with a normalized data format with a database developed. The different parts are summarized below.

The technical characteristics of marine exposure sites, in view of clear description of long-term exposure practice, should include the following aspects: the geographical location and orientation of exposure sites and facilities; the functional planning of exposure area, positioning for different types of specimens and retrieval schemes of exposed specimens; the meteorology of surrounding or regional atmosphere; the characteristics and classification of atmospheric corrosivity for the exposure sites and facilities; the hydrology of seawater, the biological growth rates (if any), and the methods to collect, record and communicate the data relevant to exposure sites. These information items constitute the environmental and technical envelope for the marine exposure practice.

In marine exposure environments, the deterioration processes are generally divided into surface processes and the inner processes in bulk concrete for different exposure zones. The inner processes are all dominated by the chloride transport via diffusion and binding while the surface processes are closely related to the different exposure zones. All environmental factors, except the seawater salinity, have notable impact on the durability processes. The binder type, together with the water to cement ratio, are the significant factors for long-term durability. The protection such as silane impregnation treatment on specimens is also of significance. The loading and the cracking, conforming to standard specifications, do not have significant impact on the durability performance of concretes.

The exposure data bear dispersion nature. The dispersion can be inherent due to the randomness of the environmental actions or the material composition, or can be artificial due to the sampling, testing methods and the interpretation models. The inherent dispersion of exposure data is discussed to show how the tools and models from engineering statistics can help to describe and handle the dispersion of data. To this purpose, the basic probability models are recalled for discrete and continuous random variables, and the distribution laws are also elaborated for the boundary values for durability parameters. Then, it is shown how to estimate the probability model parameters from observed values and how the estimate accuracy improves with number of measurements.

The sampling procedure and the test methods on exposure specimens are crucial for collecting reliable exposure data. For sampling procedure, the operation of drilling the powder samples and the necessary amount of powder sample are two important technical aspects. The drilling operation should be performed in dry condition to keep the chlorides in the samples. The appropriate particle size is probably related to the subsequent testing method while the best practice suggests a range of 160 μm to around 300 μm. The amount of powder samples is recommended to be in the range of 10–20 g depending on the maximum aggregate size. The testing methods for chloride content measurement usually differs in the extraction of chlorides from powder sample through acid solution and the subsequent treatment, and mixing with boiled/boiling water will generate the most measurable chlorides.

Finally, the normalized data presentation for exposure practice is proposed from the analysis of relevant information and collected data from active marine exposure sites. Again, this normalization serves both to complete the essential information on exposure practice to support better the interpretation of exposure data, and to facilitate the data comparison and exchange for the sake of valorizing the long-term exposure practice. The minimum collection of the normalized exposure data should include the environment, the material and the performance groups of information. On this basis, a database is built and provides basic search and plot functions for chloride profiles.

## References


1. AASHTO: Standard method of test for sampling and testing for chloride ion in concrete and concrete raw materials (T260). American Association of State Highway and Transportation Officials, Washington DC (1997)
2. Andrade, C., Elsener, B., Vennesland, Ø., Gulikers, J., Weidert, R., Raupach, M.: Test methods for on-site measurement of resistivity of concrete, RILEM TC 154-EMC: electrochemical techniques for measuring metallic corrosion. Mater. Struct. **33**, 603–611 (2000)
3. Andrade, C., Sagrera J.L., Sanyuán, M.A.: Several years study on chloride ion penetration into concrete exposed to Atlantic Ocean water. In: Testing and Modelling Chloride Ingress into Concrete – Proceedings of the 2nd International RILEM Workshop, RILEM PRO 19, Paris, France, pp. 121–134 (2000)
4. Andrade, C., Gulikers, J., Polder, R., Raupach, M.: Half-cell potential measurements – potential mapping on reinforced concrete structures, RILEM TC 154-EMC: electrochemical techniques for measuring metallic corrosion. Mater. Struct. **36**, 461–471 (2003)
5. Arya, C., Buenfeld, N.R., Newman J.B.: Assessment of simple methods of determining the free chloride ion content of cement paste. Cem. Conc. Res. **17**, 907–918 (1987)
6. ASTM: Standard test method for water-soluble chloride in mortar and concrete1 (C1218). ASTM International, West Conshohocken (2020)
7. ASTM: Standard test method for acid-soluble chloride in mortar and concrete (C1152). ASTM International, West Conshohocken (2020)
8. ASTM: Standard test method for corrosion potentials of uncoated reinforcing steel in concrete (C876). ASTM International, West Conshohocken (2022)
9. Boubitsas, D., Tang, L.P., Utgenannt, P.: Chloride ingress in concrete exposed to marine environment - field data up to 20 years' exposure. CBI report to SBUF Project 12684, Swedish Cement and Concrete Research Institute, Stockholm, Sweden (2014)
10. Castellote, M., Andrade, C.: Round-Robin test on chloride analysis in concrete - part II: analysis of water soluble chloride content, TC 178-TMC: testing and modelling chloride penetration in concrete. Mater. Struct. **34**, 589–598 (2001)
11. Castellote, M., Andrade, C.: Round-Robin test on chloride analysis in concrete - part I: analysis of total chloride content, TC 178-TMC: testing and modelling chloride penetration in concrete. Mater. Struct. **34**, 532–556 (2001)
12. Castro-Borges, P., Maldonado, L.: Initial efforts to evaluate the corrosion problems in the infrastructure of the Mexican southeast coastal zone. In: CORROSION/95, Annual Congress of NACE-International, 26–31 March 1995, Orlando, Florida, Paper No. 21 (1995)
13. Castro-Borges, P., Moreno, E.I., Genescá, J.: Influence of marine micro-climates on carbonation of reinforced concrete buildings. Cem. Concr. Res. **30**(10), 1565–1571 (2000)
14. CEN: Testing hardened concrete - part 12: determination of the carbonation resistance of concrete - accelerated carbonation method (EN 12390–12). European Committee for Standardization, Brussels (2020)

15. Chalee, W., Sasakul, T., Suwanmaneechot, P., Jaturapitakkul, C.: Utilization of rice husk–bark ash to improve the corrosion resistance of concrete under 5-year exposure in a marine environment. Cem. Concr. Compos. **37**, 47–53 (2013)
16. Chaussadent, T., Arliguie, G.: AFREM test procedures concerning chlorides in concrete: extraction and titration methods. Mater. Struct. **32**(4), 230–234 (1999)
17. CIS: Technical specification for test of chloride ion content in concrete (JGJ/T-322), China Industrial Standard. China Building Industry Publishing, Beijing (2013)
18. Fagerlund, G.: On the capillarity of concrete. Nordic Concrete Research, 1, paper 6, 20pp. (1982)
19. FHIR: Multi-dimensional diffusion of chlorides in edge and corner parts of concrete elements in marine environments, Technical report project 2018-ZJKJ-PTJS06, Guangzhou, China (2018). (in Chinese)
20. Gulikers, J.: Improved engineering model for the propagation stage of chloride-induced pitting corrosion of steel reinforcement. In: 3rd International RILEM Workshop on Testing and Modelling Chloride Ingress into Concrete, pp. 257–268. RILEM Publications SARL (2004)
21. ISO: Corrosion of metals and alloys - corrosivity of atmospheres- classification, determination and estimation (ISO 9223:2012). International Standard, Geneva, Switzerland (2012)
22. Kim, J., McCarter, W.J., Suryanto, B.: Performance assessment of reinforced concrete after long-term exposure to a marine environment. Constr. Build. Mater. **192**, 569–583 (2018)
23. Li, K.F., Zhang, D.D., Li, Q.W., Fan, Z.H.: Durability for concrete structures in marine environments of HZM project: Design, assessment and beyond. Cem. Concr. Res. **115**, 545–558 (2019)
24. Li, K.F., Zeng, J.J., Tang, L.P., et al.: Long-term field exposure of structural concretes in marine environment: state-of-the-art review by RILEM TC289-DCM. Mater. Struct. **55**, 205 (2022)
25. Li, Q.W., Li, K.F., Zhou, X.G., Zhang, Q.M., Fan, Z.H.: Model-based durability design of concrete structures in Hong Kong–Zhuhai–Macau sea link project. Struct. Saf. **53**, 1–12 (2015)
26. Mackechnie, J.R.: Predictions of reinforced concrete durability in the marine environment. Ph.D. thesis, University of Cape Town, South Africa (1995)
27. Moradllo, M.K., Sudbrink, B., Hu, Q., Aboustait, M., Tabb, B., Ley, M.T., Davis, J.M.: Using micro X-ray fluorescence to image chloride profiles in concrete. Cem. Concr. Res. **92**, 128–141 (2017)
28. Muralidharan, S., Vedalakshmia, R., Saraswathia, V., Joseph, J., Palaniswamy, N.: Studies on the aspects of chloride ion determination in different types of concrete under macro-cell corrosion conditions. Build. Environ. **40**, 1275–1281 (2005)
29. Nordtest: Concrete hardened: accelerated chloride penetration (NT Build 443). Nordtest, Espoo, Finland (1995)
30. ONNCCE: Building industry-concrete-drilled specimen and sawed beam of hardened hydraulic concrete (NMX-C-169). Mexican Standard, ONNCCE, Mexico (2009)
31. ONNCCE: Construction industry – hydraulic concrete – durability – determination of carbonation depth in hydraulic concrete – specifications and test method (NMX-C-515). Mexican Standard, ONNCCE, Mexico (2016)
32. ONNCCE: Construction industry-durability-general standard for the durability of reinforced concrete structures-criteria and specifications (NMX-C-523). Mexican Standard, ONNCCE, Mexico (2016)
33. RILEM: Recommendation of RILEM TC 178-TMC: Testing and modelling chloride penetration in concrete, analysis of water soluble chloride content in concrete. Mater. Struct. **36**, 586–588 (2002)
34. RILEM: Recommendation of RILEM TC 178-TMC: testing and modelling chloride penetration in concrete, analysis of total chloride content in concrete. Mater. Struct. **36**, 583–585 (2002)
35. RILEM: Recommendation of RILEM TC 178-TMC: testing and modelling chloride penetration in concrete, Methods for obtaining dust samples by means of grinding concrete in order to determine the chloride concentration profile. Mater. Struct. **46**, 337–434 (2013)
36. Sandberg, P., Tang, L.P., Andersen, A.: Recurrent studies of chloride ingress in uncracked marine concrete at various exposure times and elevations. Cem. Concr. Res. **28**, 1489–1503 (1998)

37. Sørensen, H., Jönsson, U., Munch-Petersen, C.: Chloride penetration into concrete—comparison of results from field exposure tests and laboratory tests. In: Proceedings of 2nd International Congress on Durability of Concrete (ICDC), 4–6 December 2014, New Delhi, India (2014)
38. Tang, L.P., Fu, Y.: A rapid technique using handheld instrument for mapping corrosion of steel in reinforced concrete. Restor. Build. Monum. **12**(5–6), 387–400 (2006)
39. Troconis de Rincón, O., Montenegro, J.C., Vera, R., et al.: Reinforced concrete durability in marine environments DURACON project: long-term exposure. Corrosion **72**(6), 824–833 (2016)
40. Vennesland, Ø., Raupach, M., Andrade, C.: Recommendation of Rilem TC 154-EMC: "Electrochemical techniques for measuring corrosion in concrete"—measurements with embedded probes. Mater. Struct. **40**, 745–758 (2007)
41. Wang, S.N., Tian, J., Fan, Z.H.: Research on theory and method of service life prediction of marine concrete structures based on exposure test and filed investigation. China Harb. Eng. **169**, 67–74 (2010). (in Chinese)

# Chapter 4
# Long-Term Field Exposure Data: Interpretation and Exploitation

**Liming Huang, Luping Tang, Kefei Li, Bowei Li, Carmen Andrade, Geng Guoqing, Rachid Cherif, and Abdelkarim Aït-Mokhtar**

## Notations List for Model Parameters

| | |
|---|---|
| $D_a$ | Apparent chloride diffusion coefficient |
| $D_a(t)$ | Time-dependent apparent diffusivity |
| $D_{a,0}$ | Apparent diffusion coefficient at time $t_0$ |
| $D_{lab}$ | Chloride migration coefficient measured in lab |
| $D_{6m}$ | The diffusion coefficient measured by the Rapid Chloride Migration (RCM) test at the age of $t_{6m}$ |
| $D_i$ | Diffusion coefficients $i$ species |
| $D_{ef}$ | Effective diffusion coefficient of chloride ion in pore solution |
| $D_1$ | Apparent chloride diffusivity for carbonated |
| $D_2$ | Apparent chloride diffusivity for un-carbonated zones respectively |
| $D_{ps}$ | Diffusivity of the ions in the bulk pore solution |

L. Huang · L. Tang (✉)
Department of Architecture and Civil Engineering, Chalmers University of Technology, Gothenburg, Sweden
e-mail: tang.luping@chalmers.se

L. Huang
e-mail: limingh@chalmers.se

K. Li · B. Li
Department of Civil Engineering, Tsinghua University, Beijing, China

C. Andrade
International Center for Numerical Methods in Engineering (CIMNE), Barcelona, Spain

G. Guoqing
Department of Civil and Environmental Engineering, National University of Singapore, Singapore, Singapore

R. Cherif · A. Aït-Mokhtar
LaSIE UMR CNRS 7356, La Rochelle University, Cedex 1 La Rochelle, France

 
K. Li et al. (eds.), *Long-Term Performance of Structural Concretes in Marine Exposure Conditions*, RILEM State-of-the-Art Reports Volume,
https://doi.org/10.1007/978-3-032-06959-7_4

| | |
|---|---|
| $D_p$ | Diffusivity of the ions in the cement paste |
| $C_{\mathrm{S}}$ | Total surface chloride content (constant) |
| $C_{\mathrm{S}}(\mathrm{t})$ | Time-dependent surface chloride content |
| $C_0$ | Initial total chloride content in concrete |
| $c_{\mathrm{s}}$ | Free chloride concentration at the surface of the concrete |
| $c_0$ | Initially free chloride in the concrete |
| $c$ | Concentration of free chloride ions |
| $c_{\mathrm{r}}$ | Reference chloride concentration |
| $k_{\mathrm{t}}$ | Effects of age on correlation between diffusion coefficient from lab and onsite |
| $k_{\mathrm{e}}$ | Environment factor |
| $k_{\mathrm{c}}$ | Curing regime factor |
| $k$ | Carbonation rate |
| $K_1$ | Ratio of diffusivity in the carbonated and non-carbonated zones |
| $K_2$ | Ratio of the carbonation rate and chloride diffusion rate |
| $K_{sq}$ | Regression slope for square root model |
| $b$ | Regression intercept for square root model |
| $n_{\mathrm{cl}}$ | Age factor of apparent diffusion coefficient |
| $p$ | Parameter for the theorem solution of Mejlbro-Poulsen model |
| $\xi_{\mathrm{D}}$ | Factor correlating $D_{6\mathrm{m}}$ to the initial apparent diffusion coefficient under exposure environment |
| $a_{\mathrm{t}}$ | Parameter for time-dependency of chloride binding |
| $f_{\mathrm{b}}, \beta_{\mathrm{b}}$ | Chloride binding constants in ClinConc |
| $E_{\mathrm{D}}$ | Activation energy chloride diffusion and binding. |
| $E_{\mathrm{b}}$ | Activation energy chloride binding |
| $f_{\mathrm{red}}$ | Redistribution factor of the bound chloride |
| $b_0, k_1, k_2$ | Redistribution factor of the bound chloride |
| $x_{\mathrm{p}}$ | Peak position of total chloride on surface zone |
| $c_i$ | Molar concentration of ionic species *i* in concrete pore solution |
| $\mu_i{}^0$ | Standard electrochemical potential of ionic species of *i* |
| $\gamma_i$ | Ionic activity coefficient of *i* |
| $z_i$ | Chemical valence of ion *i* |
| $\alpha, \beta$ | Isotherm constants in chloride binding models |
| $n_{i,s}$ | Ion stochiometric number corresponding to the solid phase |
| $\chi_s$ | Kinetic constant of the dissolution or precipitation of the solid phase |
| $K_s$ | Equilibrium constant for dissolution of the solid phase |
| $\tau$ | Tortuosity index |
| $\delta$ | Reduction factor for correlation between pore structure and ion diffusion |
| $u$ | Velocity of ions movement caused by convection |
| $S$ | Saturation degree of porous media |
| $\varphi$ | Porosity of the media |
| $\rho_{\mathrm{ef}}$ | Resistivity in saturated conditions of the concrete specimen and |
| $\rho_{\mathrm{ap}}$ | Resistivity considering the binding of the chlorides |
| $r_{\mathrm{Cl}}$ | Chloride binding factor reflected by resistivity |
| $\rho_{\mathrm{t}}$ | Resistivity at age *t* |

$\rho_0$ Resistivity at the age of the first measurement $t_0$
$k_{Cl}$ Exposure factor

## 4.1 Synthetic Analysis of Modelling

With the accumulation of extensive data from long-term exposure sites and real structures in marine environments, there is a need for robust modelling approaches to interpret this data, particularly for chloride ingress in concrete. These models provide essential insights for the durability design of reinforced concrete structures exposed to marine environments. A variety of models addressing chloride ingress have been developed, as listed in a recent review by this TC [36], with most models also covered in the earlier RILEM TC 270-CIM report [34]. In this section, we will summarize the key governing equations of these models, where notes and comments will be attached to guide users in selecting the most appropriate model based on specific requirements and contexts. An overview of models is given in Table 4.1 with special comments on the easiness of operating the model (*Complexity*), matching level of the prediction and real experimental value (*Accuracy*) and the practical level for engineering application (*Applicability*).

### *4.1.1 Models Based on Analytical Solutions to Fick's 2nd Law*

#### 4.1.1.1 *ERFC* with Constant Chloride Diffusion Coefficient

The Model of Error-function solution to Fick's 2nd law (***ERFC***) is widely used for describing chloride ingress since its introduction by Collepardi et al. [18]. The model is derived from Fick's 2nd law with constant non-steady state chloride diffusion coefficient ($D_a$), under the semi-infinite boundary condition, as

$$C(x,t) = C_0 + (C_S - C_0) \cdot \operatorname{erfc}\left(\frac{x}{2\sqrt{D_a \cdot t}}\right) \tag{4.1}$$

where $C(x,t)$ is the total chloride content at depth $x$ at exposure time $t$, $C_0$ is the initial chloride content in concrete and $C_S$ is the surface chloride content.

In this model, the key parameters are $C_S$ and $D_a$, which must be determined by curve-fitting the chloride ingress profiles from field or laboratory exposure. It has been demonstrated by numerous experimental data that this simple model can only give a satisfactory description of the chloride ingress under exposure conditions for short durations, close to the instants when the input parameters are determined.

**Table 4.1** Summary of the listed models

| Model | Input parameters | Features | | | |
|---|---|---|---|---|---|
| | | Complexity | Accuracy | Applicability | Comments |
| ***ERFC*** with constant chloride diffusion coefficient [18] | $D_a$, $C_S$, $C_0$ | *Low* | *Low* | *Low* | Model for teaching purpose |
| Empirical ***ERFC*** [34] | $D_a$, $C_S(t)$, $C_0$ | *Low* | *Medium* | *Low* | Parameters regressed from experimental results |
| ***DuraCrete*** and *fib* model [22, 23] | $D_{lab}$, $k_t$, $k_c$, $n_{cl}$, $C_S$, $C_0$ | *Medium* | *Medium* | *High* | Already used in practice |
| Mejlbro-Poulsen model [41] | $C_s(t)$, $C_0$, $\alpha$, $p$, $D_{a,0}$, $n_{cl}$ | *High* | *Medium* | *Low* | / |
| ACI Life 365 model [65] | $D_{a,0}$, $n_{cl}$, $C_S$, $C_0$ | *Medium* | *Medium* | *Medium* | Already used in practice |
| ClinConc (two-steps) model and the modified version [60, 61] | $c_s$, $c_0$, $D_{6m}$, $n_{cl}$, $a_t$, $f_b$, $\beta_b$, $E_D$, $E_b$<br>The modified version: + $f_{red}$, $c$, $b_0$, $k_1$, $k_2$ | *High* | *High* | *Medium* | Compared with other models |
| Multi-species models [33, 35, 43] | $D_i$, $c_i$, $\mu_i^0$, $\gamma_i$, $z_i$, $\alpha$, $\beta$, $n_{i,s}$, $\chi_s$, $K_s$ | *High* | *High* | *Low* | Software necessary for model application |
| Model with convection effect [46] | $\tau$, $D_{ef}$, $c_i$, $\boldsymbol{u}$, $\varphi$, $S$ | *High* | *Medium* | *Low* | |
| Square-root model [24] | $K_{sq}$, $b$, $c_r$ | *Low* | *Medium* | *Low* | Model proposed from long-term exposure data |
| Model based on electrical resistivity [1] | $\rho_{ef}$, $\rho_{ap}$, $r_{Cl}$, $\rho_t$, $\rho_0$, $k_{Cl}$, $D_{ef}$, $n_{cl}$ | *Low* | *Medium* | *High* | Model simple to use |
| Model with concurrent carbonation [37] | $D_1$, $D_2$, $k$, $K_1$, $K_2$, $c_s$, $c_0$ | *High* | *Medium* | *Medium* | Not yet applied to engineering case |

### 4.1.1.2 The Empirical *ERFC*

A simple empirical model with time-dependent diffusivity $D_a$ and time-dependent $C_S(t)$ was referred as the "false error function solution" in the State-of-the-Art Report of the RILEM TC 270-CIM [34]. This model involves using Eq. (4.1) but expressing the chloride surface content as a time-dependent parameter. Such an equation may appear to describe the data well when it is used for curve-fitting data points, but it is important to note that this equation is not a mathematically correct solution to Fick's 2nd law. Therefore, the correctness of model prediction cannot be guaranteed, especially for prediction of chloride ingress beyond the test durations from which the parameters are fitted.

#### 4.1.1.3 DuraCrete and Fib Model

The chloride diffusion coefficient in concrete varies over time in real application cases, so both the *DuraCrete* project [22] and *fib* model code [23] suggest to incorporate the time-dependency of chloride diffusivity by defining the apparent chloride diffusion coefficient as,

$$D_a(t) = k_t \cdot k_e \cdot k_c \cdot D_{\text{lab}} \cdot \left(\frac{t_0}{t}\right)^{n_{cl}} \tag{4.2}$$

where $D_{\text{lab}}$ is the chloride migration coefficient that can be measured in laboratory, e.g. by the Nordtest method NT BUILD 492 [45] at $t_0 = 28$ days ($D_{\text{RCM},0}$), or by natural diffusion test [23], the constants $k_{\text{t}}$, $k_{\text{e}}$ and $k_{\text{c}}$ account for the effects of age, environment, and curing regime, respectively; $t_0$ represents the reference period, e.g. 28 days, when $D_{\text{lab}}$ is measured; the parameter $n_{\text{cl}}$ describes the exponent factor governing the time-dependency of the apparent diffusion coefficient. Despite this adjustment, the solution to Fick's 2nd law remains the same as Eq. (4.1), with the constant $D_a$ replaced by $D_a(t)$ that represents the mean value of time-dependent apparent diffusion coefficient.

The principle outlined in the *DuraCrete* model has been adopted in guidance provided by the Concrete Society [7] and the *fib* model code for service life design [23]. In these guidelines, the factor $k_{\text{e}}$ is determined by,

$$k_e = \exp\left[b_e\left(\frac{1}{273 + T_{\text{r}}} - \frac{1}{273 + T_{\text{e}}}\right)\right] \tag{4.3}$$

where, $b_{\text{e}}$ represents the regression variable, which ranges from 3 500 to 5 500, with a mean value of 4800 and a standard deviation of 700; $T_{\text{r}}$ denotes the reference temperature in °C at which the chloride migration coefficient is measured, and $T_{\text{e}}$ represents the actual exposure temperature in °C.

#### 4.1.1.4 Mejlbro-Poulsen Model (Time-Dependent $D_{\text{a}}$ and $C_{\text{S}}$)

This mathematical model, developed by Mejlbro and Poulsen in 1996 [41] and applied in Denmark through a national project (*HETEK* project) [27], assumes the total chloride as the driving force. It considers both the surface chloride content $C_{\text{S}}$ and the apparent diffusion coefficient $D_{\text{a}}$ as time-dependent functions:

$$Cs(t) = C_0 + \alpha \cdot ((t - t_0) \cdot D_a(t))^p \tag{4.4}$$

$$D_a(t) = D_{a,0} \cdot \left(\frac{t_0}{t}\right)^{n_{cl}} \tag{4.5}$$

where $Cs(t)$ is the total surface chloride content at age of $t$, $C_0$ is the initial chloride content, $\alpha$ is a parameter for chloride adsorption (percentage/year$^p$), $p$ is a parameter for the theorem solution of above two equations, $D_{a,0}$ is the apparent diffusion coefficient at time $t_0$. Figure 4.1 shows different functions to describe the time-dependency of the chloride surface content $C_S(t)$ with various $p$ values. The general theorem solution is expressed as,

$$C(x, t) = C_0 + \alpha \cdot ((t - t_0) \cdot D_a(t))^p \psi_p(z) \tag{4.6}$$

$$z \approx \frac{x}{2\sqrt{D_a(t)t}} \tag{4.7}$$

where $\psi_p$ is a series of gamma-functions. When $p = 0$, Eq. (4.6) becomes the same form as Eq. (4.1). As it is impossible to measure $D_{a,0}$ at the time of exposure $t_0$ (no ingress profile is available for curve-fitting), the model needs some experimental data from certain exposures (e.g. one year exposure and at a later age) to estimate the values of $D_{a,0}$, $n_{cl}$, $\alpha$ and $p$. Updating from later available site data may result in different values for these empirical parameters [25, 26]. Figure 4.2 shows the correlation between $\psi_p$(z) and $z$ for $p$ ranging from −0.5 to 1 with Δp of 0.25. More detailed values of $\psi_p(z)$ can be found in Tables 1–5 of Ref. [41].

#### 4.1.1.5 ACI Life 365 Model

In North America, ACI TC-365 developed an empirical model in the early 2000s [65]. The model uses Fick's 2nd law of time-dependent diffusion coefficient as the transport function, and it considers the total chloride content $C$ as the driving force. The model focuses on the chloride ingress and is convenient to use for engineering, but the chloride ingress is greatly simplified as pure diffusion,

$$\frac{\partial C}{\partial t} = D_a(t)\frac{\partial^2 C}{\partial x^2} \tag{4.8}$$

$$D_a(t) = D_{a,0}\left(\frac{t_0}{t}\right)^{n_{cl}} \tag{4.9}$$

where $D_{a,0}$ is the apparent diffusion coefficient at the reference exposure time $t_0$ and $n_{cl}$ is a constant to describe aging effect. To prevent infinite decrease of the diffusion coefficient over time, the validity of relationship described in Eq. (4.9) is set

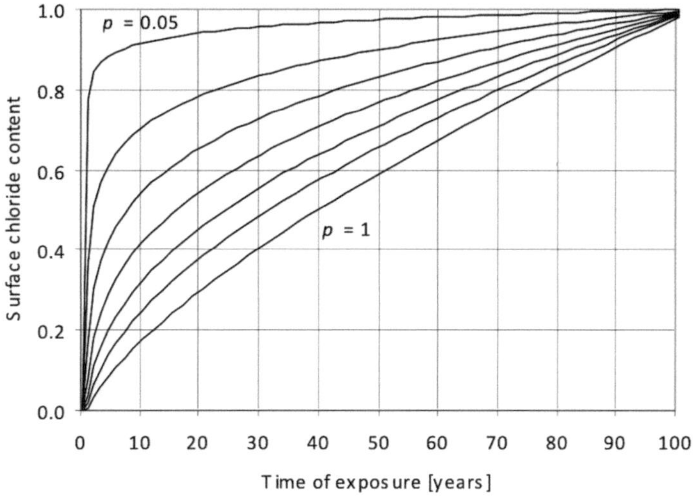


**Fig. 4.1** Different functions to describe the time-dependency of the chloride surface content $C_S(t)$ in the Mejlbro-Poulsen model. Reproduced with the permission from [41] under the terms of the Creative Commons [CC BY 4.0] © [2024] [Cementa AB]

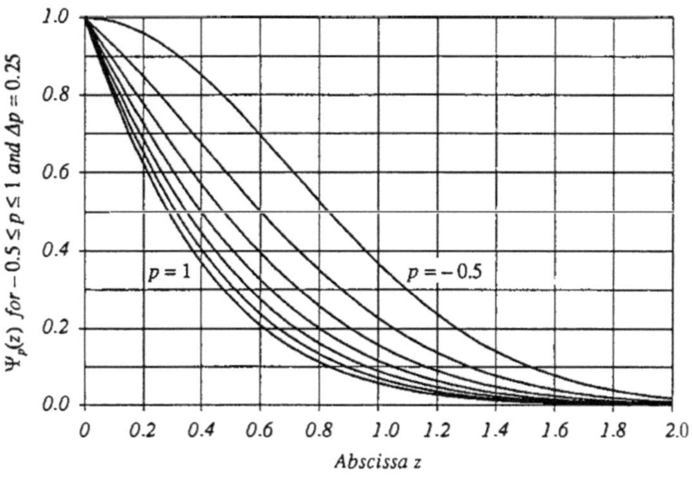


**Fig. 4.2** Value of $\psi_p(z)$ following the change of $z$ for $p$ of $-0.5$ to 1 with $\Delta p$ of 0.25. Reproduced with the permission from [41] under the terms of the Creative Commons [CC BY 4.0] © [2024] [Cementa AB]

to 30 years. After this period, the diffusion coefficient at 30 years, $D_a$(30 years), is assumed to remain constant for subsequent ages. Temperature effects on the apparent diffusion coefficient are also considered. Factors such as exposure condition, structure, and geographic location will influence the selection of the rate of chloride diffusion and the maximum surface content. Moreover, the model provides different $n_{cl}$

values for various pozzolanic material additions, making it semi-empirical. Notably, the ACI model software integrates chloride ingress, corrosion initiation and propagation, repair schedules, and life-cycle costs, offering a user-friendly tool for maintenance planning of concrete structures. The model employs a numerical approach for time integration, which requires a specialized software.

An important distinction to note is that $D_a(t)$ in Eq. (4.8) differs from $D_a(t)$ in Eqs. (4.2) and (4.4). The former represents the instantaneous diffusion coefficient as traditionally defined in Fick's law, while the latter denotes the average diffusion coefficient during the period from $t_0$ to $t$.

## 4.1.2 Models Based on Physical and Chemical Process

### 4.1.2.1 The ClinConc (Two-Steps) Model and the Modified Version

The initial version of the ClinConc model was developed in the mid-1990s [61, 63]. It comprises two primary steps: firstly, simulating the penetration of free chloride through the pore solution in concrete using a flux equation based on Fick's law, with the free chloride concentration serving as the driving potential; secondly, calculating the distribution of total chloride content in concrete by combining the mass balance equation with non-linear chloride binding. Notably, the ClinConc model employs free chloride as the driving force and incorporates non-linear chloride binding, which is a more scientifically rigorous approach to describing chloride transport in concrete compared to empirical or semi-empirical models. Subsequently, the model was reformulated in a more user-friendly format to facilitate its adoption by practicing engineers [57]. The distribution of free chloride concentration in concrete at depth $x$ is governed by the following equation:

$$\frac{c - c_0}{c_s - c_0} = 1 - \mathrm{erf}\left(\frac{x}{2\sqrt{\frac{\xi_D D_{6m}}{1-n} \cdot \left(\frac{t_{6m}}{t}\right)^n \cdot \left[\left(1 + \frac{t_{ex}}{t}\right)^{1-n} - \left(\frac{t_{ex}}{t}\right)^{1-n}\right] \cdot t}}\right) \tag{4.10}$$

where $c$, $c_s$ and $c_0$ represent the concentration of free chlorides in the pore solution at depth $x$, at the surface of the concrete, and initially in the concrete, respectively. $D_{6m}$ denotes the diffusion coefficient measured by the Rapid Chloride Migration (RCM) test at the age of 6 months $t_{6m}$, $\xi_D$ is the factor that links the laboratory-measured $D_{6m}$ to the initial apparent diffusion coefficient for the actual exposure environment. The parameter $n$ represents the age factor, which considers the decrease of diffusivity with age, $t_{ex}$ is the age of concrete at the beginning of exposure and $t$ is the duration of the exposure.

The factor $\xi_D$ can be determined by the following expression:

$$\xi_{\mathrm{D}} = \frac{\left(0.8a_{\mathrm{t}}^{2} - 2a_{\mathrm{t}} + 2.5\right)(1 + 0.59K_{\mathrm{b6m}})\exp\left[\frac{E_{\mathrm{D}}}{R}\left(\frac{1}{293} - \frac{1}{T}\right)\right]}{1 + k_{\mathrm{OH6m}}K_{\mathrm{b6m}}f_{\mathrm{b}}\beta_{\mathrm{b}}\left(\frac{c_{\mathrm{s}}}{35.45}\right)^{\beta_{\mathrm{b}}-1}\exp\left[\frac{E_{\mathrm{b}}}{R}\left(\frac{1}{T} - \frac{1}{293}\right)\right]}k_{\mathrm{D}} \tag{4.11}$$

where $a_{\mathrm{t}}$ is a parameter that describes time-dependency of chloride binding; $f_{\mathrm{b}}$ and $\beta_{\mathrm{b}}$ are chloride binding constants; $E_{\mathrm{D}}$ and $E_{\mathrm{b}}$ are activation energies of chloride diffusion and binding, respectively; $K_{\mathrm{D}}$ is the expansion factor dependent on the type of binder and water-to-binder ratio; $k_{\mathrm{OH6m}}$ and $K_{\mathrm{b6m}}$ are factors that account for the effects of hydroxide concentration in the pore solution, gel content and water-accessible porosity at the age of $t_{\mathrm{6m}}$. One of the previous studies showed that the activation energy $E_{\mathrm{D}}$ is about 42,000 J/mol and $E_{\mathrm{b}}$ is 40,000 J/mol [63].

$$k_{\mathrm{OH6m}} = \exp\left[0.59\left(1 - \frac{0.043}{[\mathrm{OH}]_{\mathrm{6m}}}\right)\right],\ K_{\mathrm{b6m}} = \frac{W_{\mathrm{gel6m}}}{1000\,\varepsilon_{\mathrm{6m}}} \tag{4.12}$$

where $[\mathrm{OH}]_{6\,\mathrm{m}}$, $W_{\mathrm{gel6m}}$ and $\varepsilon_{\mathrm{6m}}$ represent the hydroxide concentration in mol/($\mathrm{m}^3$ pore solution), gel content in kg/($\mathrm{m}^3$ concrete) and water accessible porosity at the age $t_{\mathrm{6m}}$, respectively. Based on a 10-year exposure in seawater on the Swedish west coast [58], the expansion factor $k_{\mathrm{D}}$ can be estimated with

$$k_{\mathrm{D}} = \begin{cases} 1 + 8(0.4 - w/b) + 7SF + 3800(SF \cdot FA) \cdot (SF + FA) & 0.25 \le w/b \le 0.4 \\ 1 & w/b > 0.4 \end{cases} \tag{4.13}$$

Here, *w/b* represents the water-to-binder ratio, and *SF* and *FA* denote the mass fraction of silica fume and fly ash in the total binder, respectively. This expansion factor describes the ratio of the diffusion coefficient in field concrete to that in the laboratory test.

***Determination of the total chloride content***

The total chloride content, represented by $c_{\mathrm{tot}}$ or $C$ to differentiate it from the free chloride concentration $c$, is the sum of the bound chloride $c_{\mathrm{b}}$ and free chloride $c$. However, this requires a quantitative relationship between the free and total chloride contents, known as the chloride binding isotherm. In instances where this relationship is unknown, the following expression could be employed,

$$C = \frac{\varphi \cdot (c_b + c)}{B_c} \times 100\,\text{mass \% of binder} \tag{4.14}$$

where $\varphi$ is the water-accessible porosity at the age after exposure, and $B_{\mathrm{c}}$ is the cementitious binder content (kg/$\mathrm{m}^3$ concrete), and

$$c_{\mathrm{b}} = f_t k_{\mathrm{OH6m}} K_{\mathrm{b6m}} f_{\mathrm{b}} c^{\beta_b} \exp\left[\frac{E_{\mathrm{b}}}{R}\left(\frac{1}{T} - \frac{1}{293}\right)\right] \tag{4.15}$$

where $R$ is the ideal gas constant (8.3143 J. $K^{-1}$. $mol^{-1}$), $T$ is the absolute temperature, and $f_t$ is a parameter that accounts for the time dependency of chloride binding capacity which can be determined using,

$$f_t = a_t ln\left(\frac{c - c_{in}}{c_s - c_{in}} \cdot t + 0.5\right) + 1 \tag{4.16}$$

The other parameters such as porosity, gel content and hydroxide concentration, are dependent on individual mixture proportions and can be estimated using well-established equations in the concrete handbooks, and more detailed explanation for other parameters can be found in Ref. [57].

To address the peak phenomenon in chloride profiles of on-site exposed concretes, a modified version of the ClinConc model has been proposed [62]. This modification accounts for changes in pore structure and chloride binding capacity near the surface due to the dry-wetting process experienced in splashing and atmospheric zones. Quantifying the combined effects of carbonation and leaching is very challenging for both theoretical models and experimental methods. Therefore, the empirical equation, Eq. (4.17), have been proposed to describe the redistribution of chloride binding near surface zone,

$$f_{red} = \begin{cases} \left[1 - k_1 c^{\frac{(x_{peak} - x)}{x_{peak}}} \sqrt{\frac{w}{b} - b_0}\right] k_2 \sqrt{\frac{w}{b} - b_0} \; x \le x_p \\ k_2 \sqrt{\frac{w}{b} - b_0} \; x > x_p \end{cases} \tag{4.17}$$

where $f_{red}$ is the redistribution factor of the bound chloride, $x_p$ is the peak position of total chloride on surface zone and $c$ is the concentration of free chloride predicted using the ClinConc model. The parameters $b_0$, $k_1$ and $k_2$ are constants. Differences between the ClinConc model and the modified version are illustrated in Fig. 4.3. The steps to calculate free and bound chloride content in concretes using the modified ClinConc model are as follows:

(1) Calculate the free and bound chloride content in concretes using the original ClinConc model as described in Ref. [57],
(2) Determine the redistribution factor of chloride using Eq. (4.17), and then recalculate the bound chloride profiles by multiplying the factors;
(3) Finally, calculate the total chloride profiles in concretes using the modified values as described in Ref. [57].

#### 4.1.2.2 Multi-species Models

Multi-species models can account for the complex equilibria between the pore solution and solid phases in cementitious materials [33, 35, 43]. This allows for a more accurate representation of the chemical interactions occurring within the material. Considering the free energy gradient between two points in a solution with

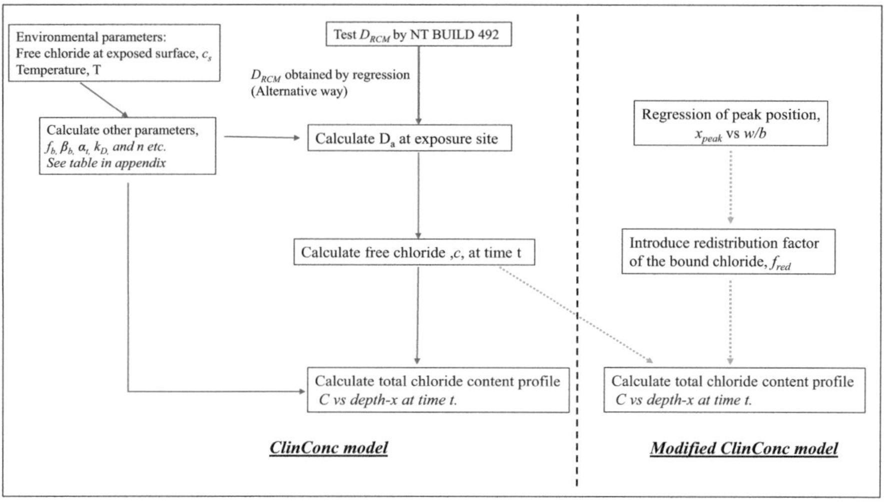


**Fig. 4.3** Flow chart for illustration of ClinConc model and the modified version of model. Reproduced from [62] under the terms of the Creative Commons [CC BY 4.0]. © [2024] [Elsevier]

different concentrations, the diffusion flux can be determined using the second law of thermodynamics [29],

$$j_i = -B^* c_i \nabla \mu_i \tag{4.18}$$

where, $\mu_i$ is the electrochemical potential. By substituting it with Eq. (4.19), the flux can be calculated from Eq. (4.20):

$$\mu_i = \mu_i^0 + z_i F \psi B + RT ln(\gamma_i c_i) \tag{4.19}$$

$$j_i = -B^* c_i \nabla \left(\mu_i^0 + z_i F \psi B^* + RT ln(\gamma_i c_i)\right) \tag{4.20}$$

where, $\mu_i{}^0$ is the standard electrochemical potential, $R$ is the ideal gas constant (8.3143 J. $K^{-1}$. $mol^{-1}$), $\gamma_i$ is the ionic activity coefficient, $T$ is the temperature (K), $F$ is Faraday's constant (9.64846 × $10^4$ C. $mol^{-1}$), $z_i$ is the chemical valence of ion *i*, and $\Psi$ (V) is the local electrical potential. The parameter $B^*$ depends on the ion diffusion coefficient $D_i$ through,

$$B^* = \frac{D_i}{RT} \tag{4.21}$$

By substituting Eq. (4.21) into Eq. (4.20), one obtains the Nernst-Planck equation,

$$j_i = -D_i \left( \nabla c_i + \frac{z_i F}{RT} c_i \nabla \psi + C_i \nabla (\ln \gamma_i) \right) \tag{4.22}$$

when using the Nernst-Planck equation to describe ion flux in the model, one may need a concurrent description of the electrical potential. The distribution of electrostatic potential can be expressed using the Poisson's equation,

$$\Delta\psi = -\frac{F}{\varepsilon_0\varepsilon_r}\sum_i z_i c_i \tag{4.23}$$

where $\varepsilon_0$ represents the vacuum dielectric constant ($8.854 \times 10^{-12}$ F/m), and $\varepsilon_r$ stands for the relative dielectric constant of water at the temperature of 298 K (78.3).

The ionic activity coefficient, $\gamma_i$, of species in a highly dilute solution is equal to 1. However, as the concentration of the solution increases, which is the case for concrete pore solution, the coefficient deviates from this value. In such solutions, two types of interactions occur. The first type involves ion–solvent interactions, where ions move through the pore solution with a tendency to drag along several solvent molecules or interact with the solid wall of pores giving rise to an electrical double layer. This phenomenon can lead to a decrease in ionic mobility. The second type involves ion-ion interactions due to the electrical charge carried by each species. To calculate the ionic activity coefficient, transport models usually use the Debye-Hückel equation, which relates the ionic strength ($I$) to the activity coefficient,

$$ln\gamma_i = -Az_i^2\sqrt{I} \tag{4.24}$$

where $A$ is the Debye-Hückel constant.

The basic assumption in the Debye-Hückel model is that a cloud of ions with opposite charges surrounds a central ion. All particles are treated as point charges (having zero radius). This model represents long-range interactions between ions. Later, the model was modified to account for the finite radius of the central ions ($a_i$). The extended Debye-Hückel model is given by,

$$ln\gamma_i = \frac{-Az_i^2\sqrt{I}}{1 + Ba_i\sqrt{I}} \tag{4.25}$$

where the parameter $B$ is the Debye-Hückel constant. The calculation of $A$ and $B$ was given in the work of [53].

Several models have been developed, taking into account both short- and long-range interactions [38, 48]. These models are generally composed of two parts: the first part based on the Debye-Hückel law, representing long-range interactions, and the second part considering ionic strength, solvation, and short-range interactions. The most well-known model is the Davies model,

$$ln\gamma_i = \frac{-Az_i^2\sqrt{I}}{1 + Ba_i\sqrt{I}} + \frac{CAz_i^2\sqrt{I}}{\sqrt{1000}} \tag{4.26}$$

where $C = 0.2$. Based on comparisons with experimental results of the ionic activity, [53] observed that the Pitzer model for calculating the ionic activity coefficient best represents reality. It is applicable for highly concentrated solutions with ionic strengths up to 1 mol $l^{-1}$. However, the Davies model is suitable for ionic strengths below 0.5 mol $L^{-1}$.

In order to simulate the time evolution of chloride concentration during transport, the Nernst-Planck equation is incorporated into the mass balance equation,

$$\frac{\partial c_i}{\partial t} + \frac{\partial C_{Cl,b}}{\partial t} = \nabla.\left(D_i \nabla c_i + D_i \frac{z_i F}{RT} c_i \nabla \psi + D_i c_i \nabla (\ln \gamma_i)\right) \tag{4.27}$$

where the term $C_{Cl,b}$ is the concentration of chlorides bound to solid matrix of concrete. The bound chlorides can be calculated through the linear isotherm, the Freundlich isotherm and the Langmuir isotherm [8, 51, 60] as follows,

$$C_{Cl,b} = \alpha c_{Cl},\ C_{Cl,b} = \alpha c_{Cl}^{\beta},\ C_{Cl,b} = \frac{\alpha c_{Cl}}{1 + \beta c_{Cl}} \tag{4.28}$$

where the terms $\alpha$ (−) and $\beta$ (−) are the chloride isotherm constants, fitted from chloride binding measurements [50].

More recently, the thermodynamic equilibria-based modelling of reactive chloride transport in blended cementitious materials has been developed. In addition to ion-ion interactions, the model incorporates liquid–solid interactions through a thermodynamic approach, specifically addressing the dissolution and precipitation of minerals within the pores [13]. The thermodynamic coupling with chloride transport is given by a mass exchange term ($f_i$) that is added to the mass balance equation,

$$\frac{\partial \varphi c_i}{\partial t} = -\nabla J_i + f_i = -\nabla J_i \pm n_{i,s} \times r_s \tag{4.29}$$

The mass exchange term $r_S$ considers the ion concentration gain or loss in the pore solution due to the dissolution or precipitation of solid phases during the chloride transport. It is calculated from a thermodynamic approach using the mass action law and the rate $r_s$ of dissolution or precipitation of the solid phase writes,

$$r_s = \chi_s\left(1 - K_s^{-1} \prod_{i=1}^{N_i} \{c_i\}^n\right) \tag{4.30}$$

where, $n_{i,s}$ is the ion stochiometric number corresponding to the solid phase. $\chi_s$ (mol $m^{-2}$ $s^{-1}$) is the kinetic constant of the dissolution or precipitation of the solid phase, $K_s$ (−) is the equilibrium constant of the solid phase.

#### 4.1.2.3 Model with Convection Effect

In marine environments, chloride ingress through pure diffusion alone is rare. Civil structures face continuous wetting and drying from tidal and splash actions, making cyclic wetting and drying conditions crucial in simulating corrosion. When structures are submerged or below the low tidal level, corrosion is limited due to reduced oxygen supply underwater [56]. When structures are exposed to drying and wetting, the rate of chloride penetration is accelerated by convection during wetting, while exposure to atmospheric air during drying leads to an inward flow of chloride ions, driven by the high concentration gradient at the surface caused by moisture depletion. The flux of chloride ions through concrete pores, accounting for both diffusion and convection, is expressed as [46],

$$J_{Cl} = -\frac{\varphi S \delta}{\tau} D_{ef} \frac{\partial c_i}{\partial x} + \varphi S \boldsymbol{u} c_i \tag{4.31}$$

where $J_{Cl}$ is chloride ion flux, $\tau$ is tortuosity accounting for complex micro-pore structure, $\delta$ is a reduction factor to describe the interaction between pore structure and ion transport, $D_{ef}$ is the effective diffusion coefficient of chlorides in pore solution, $c_i$ is the concentration of chlorides in pore solution, $S$ is the degree of saturation of porous media, $\boldsymbol{u}$ is the flow of pore solution caused by convection, and $\varphi$ is the total porosity.

### *4.1.3 Other Models*

#### 4.1.3.1 Square Root Model

This empirical model, using long-term chloride exposure data, dates back to 1985 [49] when a simple square-root of time model was introduced for the ingress of a specific *reference chloride content*, typically the critical chloride content for corrosion initiation, as

$$X_{cr}(t) = K_{sq}\sqrt{t} + b \tag{4.32}$$

where $t$ is the exposure time, $X_{cr}(t)$ is the depth with reference chloride content ($c_r$) for initiation of corrosion at $t$, and $K_{sq}$ and $b$ are empirical constants.

A recent investigation by [24] assessed the applicability and limitations of this model. The study involved analyzing field data from various exposure conditions such as SUB, TID, SPL, ATM and inland deicing-salt exposure across different geographical locations, encompassing a range of reference concentrations for chlorides. A total of 237 combinations of concrete, exposure conditions, and reference concentrations were examined. As Fig. 4.4 shows, it was discovered that the parameters defining the linear relationship were influenced by the chosen critical concentrations and concrete

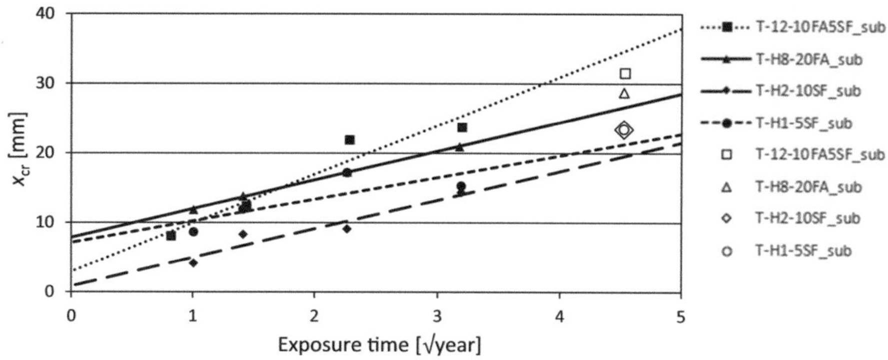


**Fig. 4.4** Predicted versus measured chloride ingress ($C_r = 0.75$) at 20 years for four submerged concretes exposed at Träslövsläge Field Exposure Site. Prediction was made from data up to 10 years exposure (shown with filled markers). Reproduced from [24] under the terms of the Creative Commons [CC BY 4.0], © [2024] [Springer]

compositions. Moreover, the slope of the relationship between $X_{cr}(t)$ and square root of $t$, referred to as the ingress parameter, was found to be dependent on the exposure conditions.

One key limitation is that it requires long-term collection of onsite data for curve fitting. This means that to apply the model accurately, extensive field data over extended periods of time are needed to calibrate the empirical constants $K_{sq}$ and $b$. Consequently, the model is more suited for predicting the remaining service life of structures based on existing exposure data. However, once the fundamental factors are clarified behind these empirical parameters, this square-root law can be expected to apply to predictive service life design of concrete structures.

#### 4.1.3.2 Model Based on Electrical Resistivity

A model based on the Einstein's relation between diffusivity and resistivity has been proposed as follow [1, 2],

$$D_{ef} = \frac{K}{\rho_{Cl}} \tag{4.33}$$

where $K$ is a constant depending on the solution concentration and adopting a value of $1 \times 10^{-4}$, $\rho_{Cl}$ is the electrical resistivity in Ω·cm and $D_{ef}$ is the diffusion coefficient in aqueous solution, also named as effective diffusion coefficient. In other terms, the effective diffusion coefficient $D_{ef}$ in Fick's laws can be solved by the above equation through the resistivity.

Although apparently simple, this expression cannot be applied to concrete without taking several assumptions. The most important facts are that concrete is a porous material, the diffusion coefficient obtained from a chloride profile is the "apparent" or "non-stationary" coefficient while in Einstein relation the formula is developed for aqueous solutions without considering any chloride binding for cement pastes in concrete. Another limitation with Eq. (4.33) is that it does not contain the time and therefore it is needed to introduce it in any of the Fick's formulas where the time is explicit. The basic one is the so named "square root law" that indicates the quadratic relation between time and chloride ingress depth,

$$x = \sqrt{2D_{ef}t} = V_{Cl}\sqrt{t} \tag{4.34}$$

where $x$ is the penetration depth of the chloride threshold for steel corrosion, $t$ is the time, $D_{ef}$ is the effective diffusion coefficient in pore solution and $V_{\mathrm{Cl}}$ is the rate of chloride penetration.

To overcome the aforementioned limitations, several assumptions were taken to modify the original Einstein expression.

1. Porous media: The resistivity adopts that of the concrete and not that of the pore solution, using the value $\rho_{\mathrm{ef}}$ at 28 days after curing the specimens in wet conditions.
2. Reaction/binding effect: a binding factor, $r_{\mathrm{Cl}}$, is introduced by an apparent diffusion coefficient, $D_{\mathrm{a}}$, considering the binding of chlorides with cement hydrated phases,

$$r_{Cl} = \frac{D_a}{D_{ef}} = \frac{\rho_{ap}}{\rho_{ef}} \tag{4.35}$$

   where $\rho_{ef}$ is the resistivity in the saturated concrete specimen and $\rho_{ap}$ is the corresponding resistivity considering the binding of chlorides. The factor $r_{\mathrm{Cl}}$ can be obtained from a two-chamber test that determines both diffusion coefficients [11]. Some results on the reaction factor are given in Table 4.2.
3. Introducing an ageing factor, $q$ [3] as

$$\rho_t = \rho_0 \left(\frac{t}{t_0}\right)^q \tag{4.36}$$

   where $\rho_{\mathrm{t}}$ is resistivity at any age $t$, $\rho_0$ is resistivity at the age of the first measurement $t_0$, and $q$ means aging factor. The relation between $q$ and the aging factor $n_{\mathrm{Cl}}$ of the diffusion coefficient gives $q = 0.8n_{\mathrm{Cl}}$. Values of the ageing factor are shown in Table 4.3.

**Table 4.2** Reaction factors of chlorides $r_{\mathrm{Cl}}$ for 3 types of cement

| Cement | $r_{\mathrm{Cl}}$ | Standard deviation |
|---|---|---|
| CEM I | 1.9 | 1.3 |
| CEM I + silica fume | 1.5 | 0.5 |
| CEM IIA (with pozzolan and fly ash, in $\leq 20\%$) | 3.0 | 2.1 |

**Table 4.3** Some values for the ageing factor $q$

| Cement | $q$ | Standard deviation |
|---|---|---|
| I | 0.22 | 0.01 |
| II/A -P | 0.37 | 0.06 |
| II/A-V | 0.57 | 0.08 |

**Table 4.4** Values of exposure factor $k_{Cl}$ for the exposure classification of EN206

| Exposure class | $k_{Cl}$ ($cm^3\Omega$/year) |
|---|---|
| XD1 / XS1 (distance to the coast line > 500 m) | 5,000 |
| XD1 / XS1 (distance to the coast line within 500 m) | 10,000 |
| XD2 / XS2 | 17,000 |
| XD3 / XS3 | 25,000 |

4. Exposure factor $k_{Cl}$: the factor depends on the concentration of chlorides. For the exposure classes in European Standard EN 206, the values are given in Table 4.4.

In summary, the proposed model for the initiation time $t_l$ of corrosion due to chlorides is then expressed by,

$$t_l = \frac{x^2.\rho_{ef}\left(\frac{t}{t_0}\right)^q}{2k_{Cl}} \cdot r_{Cl} \tag{4.37}$$

Example of application: Assuming a case where the design life is 100 years and the minimum cover depth is 5 cm, which would be the resistivity at 28d if the cement type has 18% fly ash with an aging factor 0.3 and the exposure class is the submerged zone? Using Eq. (4.37) the required effective resistivity at 28 days would be:

$$\rho_{ef} = \frac{t_l \cdot K_l}{2x^2 \cdot \left(\frac{t}{t_0}\right)^q \cdot r_{Cl}} = \frac{100 \cdot 17000}{2.25 \cdot \left(\frac{25}{0.0767}\right)^{0.3} \cdot 3} = 19.97\Omega\ m \tag{4.38}$$

#### 4.1.3.3 Chloride Ingress with Concurrent Carbonation

An analytical model for chloride ingress and concurrent carbonation was provided by [37]. The chloride ingress and the carbonation are represented in the Fig. 4.5: the chlorides diffuse across the concrete containing the carbonated and uncarbonated zones, and the carbonation occurs simultaneously with the chloride transport, idealized as a reaction front $x = L_C$.

$$\begin{aligned} \frac{\partial c_{Cl}}{\partial t} &= D_1 \frac{\partial^2 c_{Cl}}{\partial x^2}\ 0 \le x \le L_C \\ \frac{\partial c_{Cl}}{\partial t} &= D_2 \frac{\partial^2 c_{Cl}}{\partial x^2}\ x > L_C, L_C = k\sqrt{t} \end{aligned} \tag{4.39}$$

where $D_1$ and $D_2$ stand for the apparent chloride diffusivity for carbonated and uncarbonated zones respectively, and the coefficient k the carbonation rate (m/s$^{1/2}$ or mm/a$^{1/2}$) [47]. The initial and boundary conditions are,

$$c_{\mathrm{Cl}}(t=0)=c_0,\ c_{\mathrm{Cl}}(x=0,t>0)=c_{\mathrm{S}}>c_0$$
$$c_{\mathrm{Cl}}\big|_{x=L_{\mathrm{C}}^-}=c_{\mathrm{Cl}}\big|_{x=L_{\mathrm{C}}^+},\ D_1\frac{\partial c_{\mathrm{Cl}}}{\partial x}\bigg|_{x=L_{\mathrm{C}}^-}=D_2\frac{\partial c_{\mathrm{Cl}}}{\partial x}\bigg|_{x=L_{\mathrm{C}}^+} \tag{4.40}$$

The above two equations constitute a moving boundary problem (MBP) with constant initial and boundary chloride concentrations. Some simplifications are made here allowing for the analytical solution: (1) the pore saturation degree is assumed to be constant; (2) the apparent diffusivities are also constant. The solution of MBP with constant diffusivities takes the form,

$$\begin{cases} c_1=\frac{c_{\mathrm{Cl}}-c_0}{c_{\mathrm{S}}-c_0}=A_1\mathrm{erfc}\left(\frac{x}{2\sqrt{D_1t}}\right)+A_2\ x\le k\sqrt{t} \\ c_2=\frac{c_{\mathrm{Cl}}-c_0}{c_{\mathrm{S}}-c_0}=A_3\mathrm{erfc}\left(\frac{x}{2\sqrt{D_2t}}\right)\ x>k\sqrt{t} \end{cases} \tag{4.41}$$

with the constants,

$$A_{1,2,3}=A_{1,2,3}(K_1,K_2)\ \text{and}\ K_1=\sqrt{\frac{D_1}{D_2}},\ K_2=\frac{k}{2\sqrt{D_2}} \tag{4.42}$$

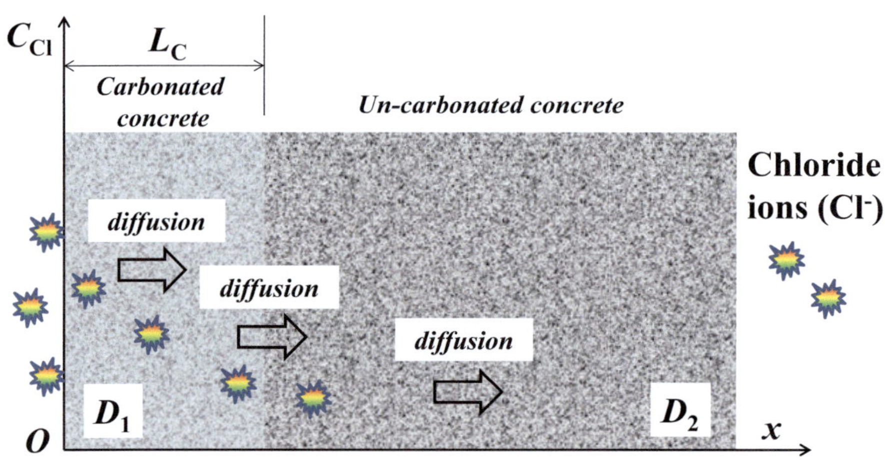


**Fig. 4.5** Chloride ions accumulate on concrete surface, and diffuse respectively across the carbonated zone, the carbonation front and into the uncarbonated zone. Reproduced from [37], with the permission © [2024] [Elsevier]

The detailed expression of $A_{1,2,3}$ can refer to Appendix A in [37]. The parameters $K_1, K_2$ are dimensionless kinetics parameters: $K_1$ scales the diffusivity ratio between the carbonated and non-carbonated zones while $K_2$ stands for the ratio between the carbonation rate and chloride diffusion rate.

The $K_1$ values being larger or smaller than 1.0 will determine whether the chloride diffusion is to be accelerated or decelerated by the carbonation. The general impact of carbonation will usually give $K_1 > 1.0$, meaning the carbonation is to accelerate the chloride diffusion, and the durability design against the chloride ingress can be unsafe without taking the carbonation into account. Once $K_1$ value is fixed, the magnitude of carbonation impact on chloride ingress is determined by the magnitude of $K_2$ values. The parameter $K_2$ scales the ratio between the carbonation rate and chloride diffusion, closely related to the pore saturation degree of concrete. The significant range is judged as $K_2 > 0.01$, i.e. pore saturation below 0.6 for OPC concretes. This range bears some general meaning: only in relatively dry environment can the carbonation have significant impact on the chloride ingress.

A comprehensive transport-reaction model of concurrent carbonation and chloride ingress into concrete was proposed by [70]. The multi-species nature of carbonation is described by the dissociation and dissolution–precipitation equilibria in pore solution. The dissolution of C-S-H is considered through the relation between the Ca/Si ratio and the portlandite saturation index. Chloride adsorption laws are regressed for C-S-H with different Ca/Si ratios. The pore transport takes into account $CO_2$ diffusion, aqueous species transport, moisture transport and the local electrical field.

The simulations confirm that the "carbonation front" holds for natural carbonation. For the retained parameters, the overall corrosion risk, through indicator $Cl^-/OH^-$, is globally increased by carbonation, and the risk depth ($Cl^-/OH^-$ above 1.0) is 20% more if carbonation is considered. The impact of concurrent carbonation on chloride ingress seems to be negligible in extreme RH conditions: too low RH will limit chloride ingress behind the carbonation front and too high RH will keep the carbonation front very near to surface. Accordingly, the significant range will be in the medium RH levels. Using the assumptions made for SCMs, the concretes incorporating SCMs, fly ash or slag, behave differently to the concurrent carbonation and chloride ingress.

#### 4.1.3.4 Machine Learning Models

Chloride ingress in concrete is affected by many factors such as concrete composition, pore structure and exposure conditions. Traditional chloride ingress models are usually incapable of considering all the influential factors. Simplification assumptions are thus inevitable, which may affect the prediction capacity of models [12]. Recently proposed numerical models considering physical, chemical and electrochemical processes of ionic diffusion showed reasonably accurate results while possessing challenges in practical engineering design [19, 42].

Machine learning (ML) is a potential alternative to address this challenge by learning from existing experimental datasets to discern unknown patterns. A comprehensive training dataset and choices of input variables (influencing factors) are often essential to obtain a reliable and robust ML model. To optimize the ML model for simplicity and adaptability, feature selection is usually incorporated to determine the number of input variables based on the importance of each feature and level of correlation between variables from the initial input data. Before model training, the raw data is usually processed. Outliers should be removed, and missing input features can be either discarded or predicted by learning algorithm.

Regression and classification are two algorithm categories within the commonly used supervised learning which predicts the target output for unknown input in the future after being trained with the input examples and desired outputs. Tran [67] demonstrated the effectiveness of Gradient Boosting and quantified the effect of supplementary cementitious material (fly ash, GGBFS and silica fume) on chloride diffusion coefficient by individual conditional expectation. Hosseinzadeh et al. [30] employed an experimental database with 1073 datasets including various types of concrete and a novel data cleaning technique (feed-forward back propagation network method) to come up with a universal machine learning model and they indicated the high performance of Extreme Gradient Boost in regression model for predicting the non-steady-state migration coefficients and classification model for identifying the chloride penetration resistance. Cai et al. [9] showed that an ensemble weighted-voting based machine learning model exhibited a better predication accuracy in surface chloride concentration than standalone models, viz., linear regression, Gaussian process regression, support vector machine, random forest and multilayer perceptron artificial neural network. The performance of different machine learning models depends on the specifics of the problems as shown in recent studies. Other generally used algorithms include neural networks, gene expression programming, and support vector machines and they are proved to be effective in predicting the chloride resistance for their calibration [20, 39].

## 4.2 Review on Benchmarking of Models

Since 21st century, three major actions of benchmarking were, to the knowledge of authors, organized at European or international levels to evaluate various models for prediction of chloride ingress in concrete under marine as well as de-icing salt road environments. This section summarizes the results from those benchmarks.

### *4.2.1 The 1st Benchmarking*

The first benchmarking was initiated in conjunction with the Nordic (and *fib* TG) seminar in May 2001 [44], utilizing the data of chloride ingress in one type of concrete from the exposure site up to two-year exposure at Träslövsläge Harbor, the western coast of Sweden, cf. Fig. 1.3. The concrete numbered "H4" (Swedish sulphate resistant, low alkali Portland cement corresponding to CEM I 42.5 N—SR 3 MH/LA, with 5% replacement of silica fume in the form of slurry, with a water-binder ratio of 0.40) was chosen in the case study, because during that time the concrete blended with 5–10% silica fume was considered as a good solution to durability of concrete exposed to chloride environments. In the benchmarking some 17 model developers and model users from nine counties (seven from Europe, one from Japan and one from Canada) were invited. The participants were supplied with the following information:

- Concrete mix proportions;
- Chloride migration coefficients measured by the RCM test (NT BUILD 492) at 0.5, 1, 1.3 and 2 years;
- Apparent chloride diffusion coefficient measured by the immersion test (NT BUILD 443) at 0.5 years;
- Chloride profiles measured after exposure for 0.6, 1 and 2 years; and
- Average annual chloride concentration in the seawater and temperature in the air at the exposure site.

They were asked to predict the chloride profiles after an exposure of 30, 50 and 100 years, and the time to initiation of corrosion, in the given concrete under two exposure environments, i.e.,

(a) marine environment with average annual chloride concentration $[Cl^-] = 14 \pm 4$ g/l in the seawater and average annual air temperature $T = 11 \pm 9$ °C and
(b) de-icing salt road environment at a distance from the road lane of 3 m, with an average of totally 0.15 kg/m$^2$ NaCl spread during five "winter" months each year.

Some eight modelers carried out the predictions and reported their prediction data mainly for the marine environment. Figures 4.6 and 4.7 summarize those predicted chloride profiles for SUB and SPL zones, respectively. It can be seen from the predicted profiles that the scatter in the predictions was large, even for SUB zone where the exposure condition is relatively constant. For SPL zone the scatter in the predicted profiles after 100 years was extremely large. It was believed that the background exposure data or input parameters for each prediction model are responsible for this large scatter. During that period no more data from the field exposure was available for verification of the predicted results.

### 4.2.2 *The 2nd Benchmarking*

The 2nd benchmarking was carried out in the middle of 2000s under the EU project ChlorTest, led by IETcc in Spain, cooperated by SP and LTH in Sweden as well as EDF in France, which was simultaneously as part of the tasks of the RILEM TC

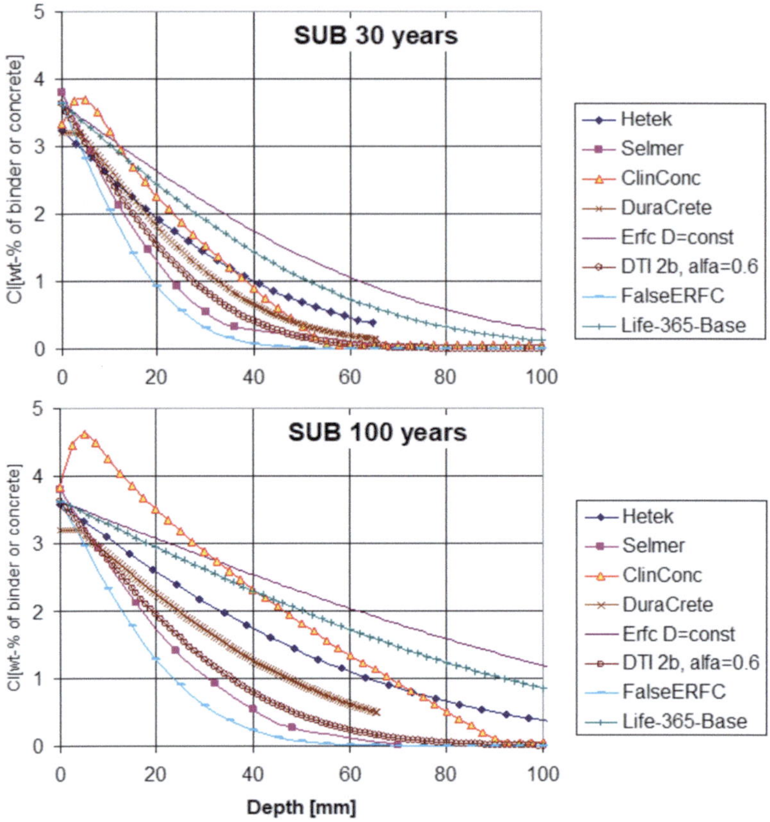


**Fig. 4.6** Predicted chloride profiles in the submerged zone of a marine environment. Reproduced with permission from [44]

178-TMC: "Testing and Modelling Chloride Penetration in Concrete" chaired by Andrade and Kropp [4], with an objective to investigate the ability of various models for prediction of chloride ingress using laboratory test results as input and in-field performance data as verification. Some 13 cases of data from the field exposure site in Sweden and 9 cases of data from Spain were selected for the benchmarking. These data included the mix proportions and properties of concrete and the environmental conditions, as well as chloride ingress profiles after a short-term (0.4–1 year for the Swedish field data and 6–8 years for the Spanish field data) exposure in the marine environments. For the Swedish cases, the chloride migration coefficients were also supplied. The model developers were asked to make predictions of chloride profiles after 10-year and 100-year exposure with their models. A probabilistic approach was used for analysis of the predicted results. The following procedures were involved in this benchmarking:

- Establishment of criteria for the benchmarking,

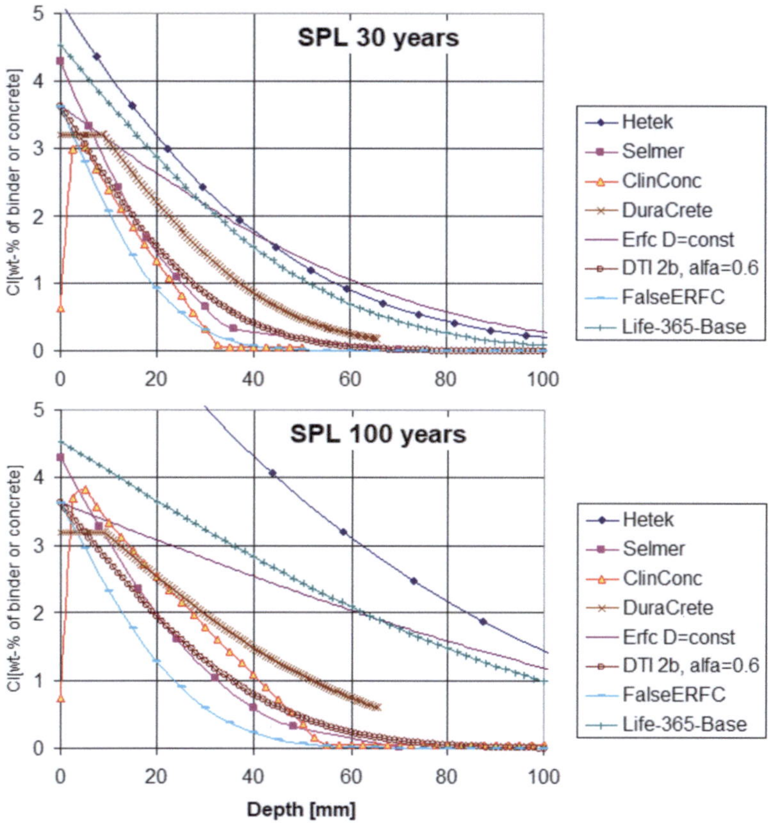


**Fig. 4.7** Predicted chloride profiles in the splash zone of a marine environment. Reproduced with permission from [44]

- Selection of profiles from the database,
- Selection of the model producers with worldwide reputation,
- Submission of the documentation to the modelers,
- Collection of results and application of benchmarking criteria, and
- Discussion on results obtained and suggestions for further actions.

In this benchmarking a methodology based on the area between the measured and the predicted profiles was developed, as shown in Fig. 4.8. The predicted profile from each model was compared with the measured profile by comparing the overlapped areas *A1* and *A2* between the depth of $x_1$ and $x_2$. The following equations are used for evaluating these overlapped areas,

$$S1 = |A1| + |A2|, S2 = A1 + A2 \tag{4.43}$$

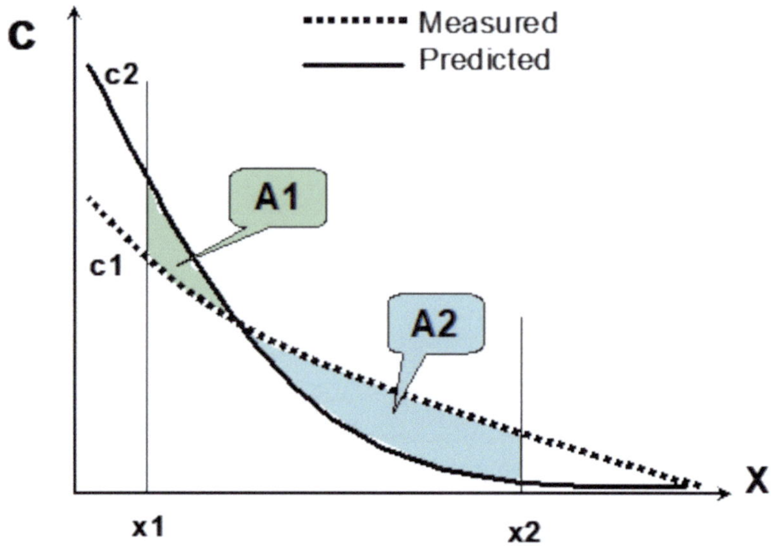


**Fig. 4.8** Methodology for evaluating the predicted chloride profiles with the measured ones. Reproduced with permission from [5, 64]

where *S*1 gives the information about how close the predicted profile is to the measured data, and *S*2 shows how higher/lower the predicted profile is than the real profile.

Some 16 models were considered in the 2nd benchmarking. The models were coded for keeping anonymous. An example of some predicted profiles for Swedish case 1 (CEM I concrete with *w/b* 0.4, SUB zone) from this benchmarking is shown in Fig. 4.9. The evaluated values of S1 and S2 are shown in Fig. 4.10. Among all the models two of them (Models 5 and 9) revealed relatively better prediction in this case of 10-year exposure. It can be seen from Fig. 4.9 that the scatter in the predicted profiles is large. The predictions for the other cases also scattered very much. More detailed description of this benchmarking can be found in [5, 64].

### *4.2.3 The 3rd Benchmarking*

The 3rd benchmarking was carried out in few years ago (2017–2018) by RILEM TC 270-CIM. Some 14 models including conventional and newly developed, physically and chemically based, models were involved in this benchmarking. Three cases were selected for benchmarking. One case for marine environment was based on the exposure site at Träslövsläge harbour, western coast of Sweden, and other two cases for deicing salt road and road/tunnel environments were based on the Bondaduz highway and the Naxberg-tunnel field experiments in Switzerland. For Case 1, the

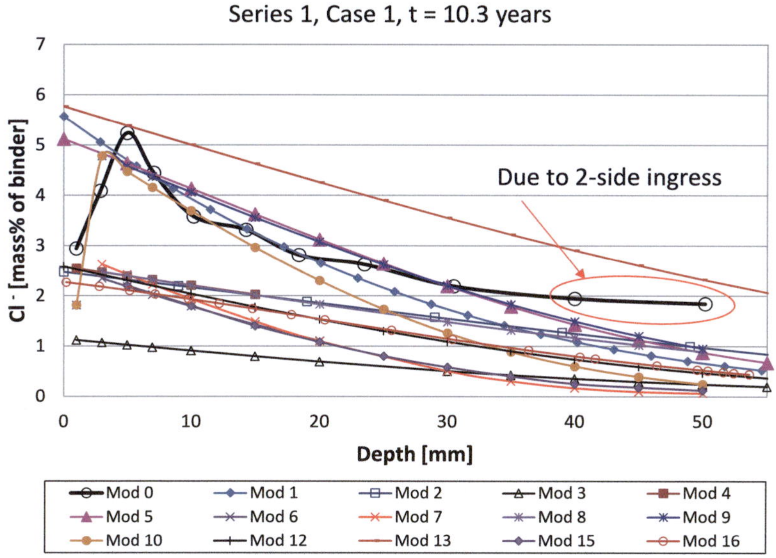


**Fig. 4.9** Predicted profiles in a CEM I concrete with a w/c of 0.40, constantly submerged in the seawater after exposed for 10.3 years. Mod 0 shows the measured data. Reproduced with permission from [5, 64]

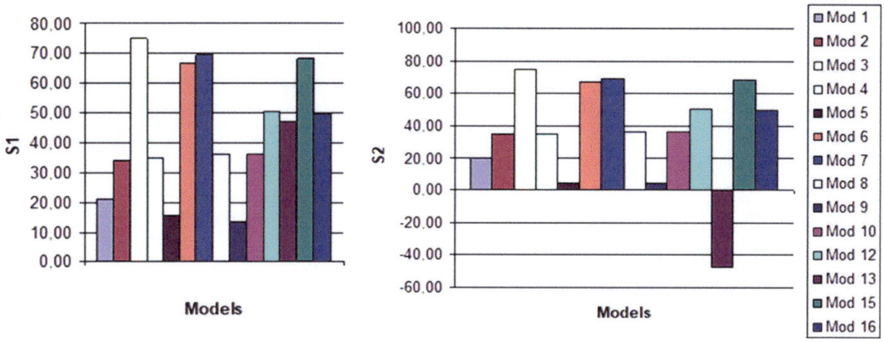


**Fig. 4.10** The evaluated values of S1 and S2 for the predicted profiles in Fig. 4.9. Reproduced with permission from [5, 64]

information of the similar type of concrete as used in the 1st benchmarking, i.e. the concrete with 90% CEM I + 5% silica fume with $w/b = 0.4$, was supplied for calibration of various models which were then used to predict chloride profiles after 5, 10, 20, 50 and 100 years of exposure. The profiles measured from the field exposure site after 5, 10 and 20 years were used for verification of the predicted profiles. In this benchmarking, a method based on the root mean square deviation (RMSD) was applied for evaluation of the predicted results. Figure 4.11 shows the predicted and measured profiles and Fig. 4.12 shows the RMSD values of various models. It was

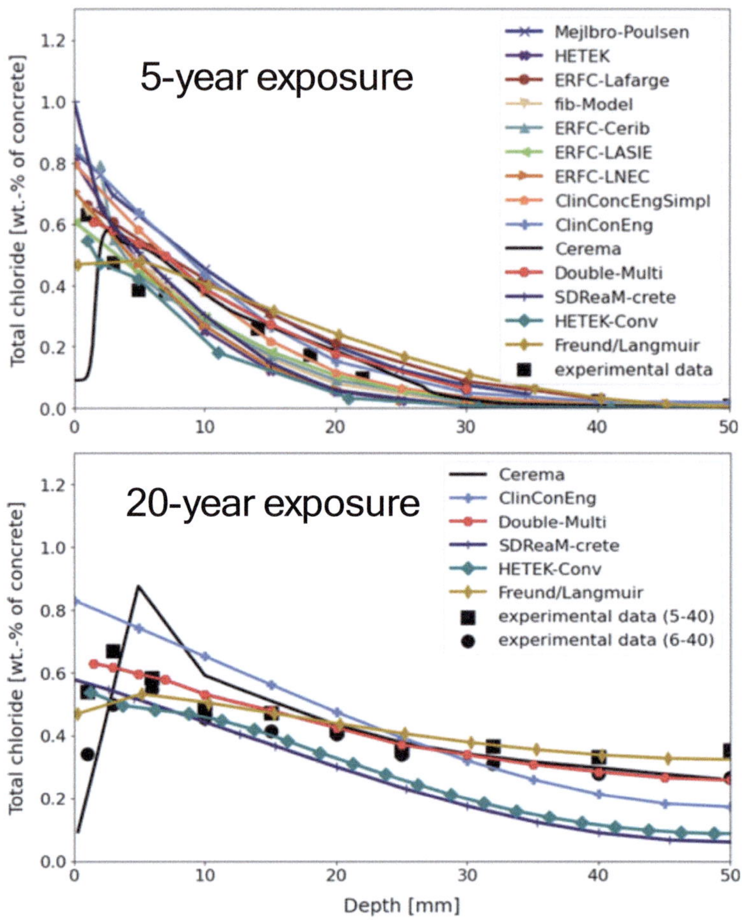


**Fig. 4.11** Predicted and measured profiles in a concrete with 95% CEM I + 5% silica fume, w/b 0.40, constantly submerged in the seawater after exposed for 5 and 20 years, adopted from [34] with permission © [2024] [Springer]

found that the scatters among predicted results after 100 years differ significantly in the range of depth at 0–10 mm, as shown in Fig. 4.13.

## 4.3 New Indicators and Interpretations

The chloride diffusion coefficient and microstructure of exposed concrete undergo continuous changes due to the ongoing hydration and environmental influences. When evaluating the performance of reinforced concrete in exposure conditions,

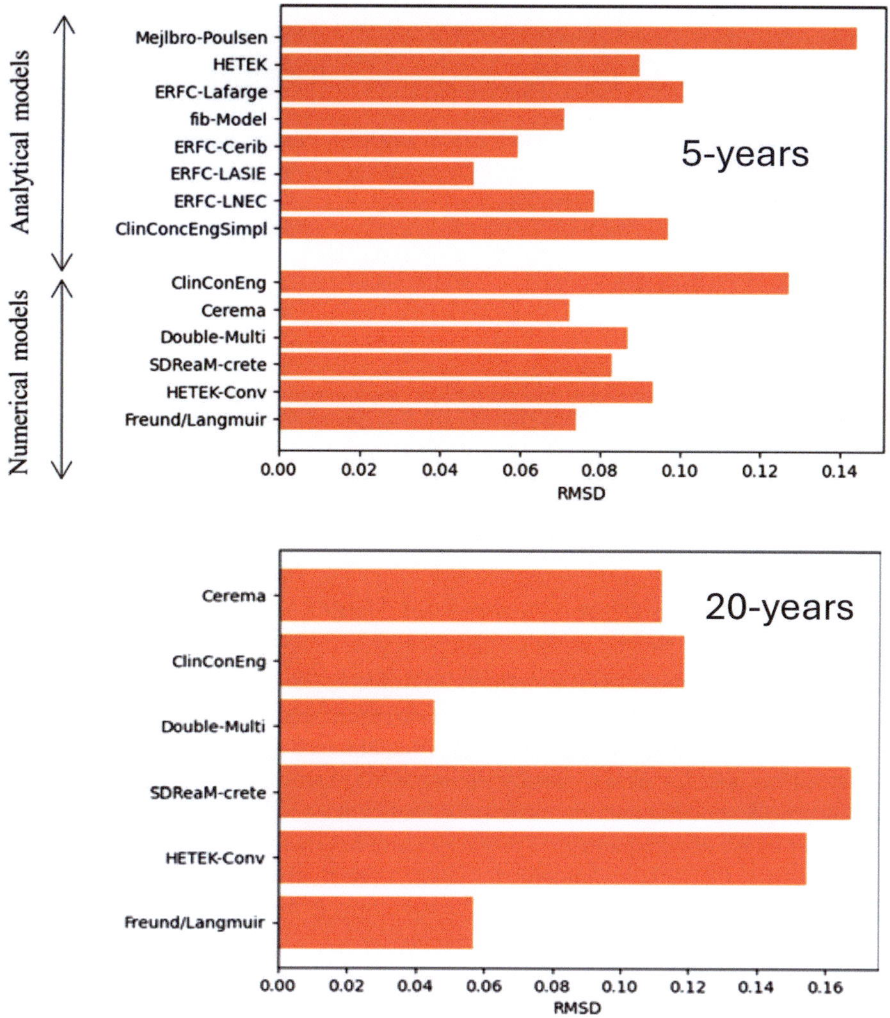


**Fig. 4.12** The RMSD values for the predicted profiles in Fig. 4.11, adopted from [34] with permission © [2024] [Springer]

it is challenging to measure directly the real-time diffusion coefficient, pore structure, or chloride binding capacity of onsite concretes. Moving beyond conventional modelling approaches, this section introduces alternative indicators, such as resistivity, the formation factor/tortuosity, and other factors generated from chloride profiles. These indicators complement existing models and provide practical metrics for engineering applications, thereby improving the assessment of RC structure durability and aiding in material selection and structural design decisions.

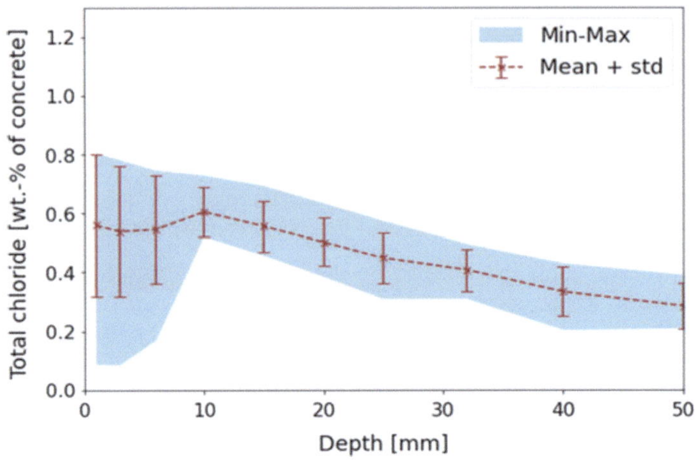


**Fig. 4.13** Predicted profiles in the 3rd benchmarking for 100-year exposure, adopted from [34] with permission © [2024] [Springer]

### *4.3.1 Parameters from Square-Root Model*

The square root law can serve to scale the chloride ingress depth in function of time following Eq. (4.32). In this equation, the rate coefficient $K_{sq}$ can be obtained from the corresponding long-term exposure data, i.e. chloride profiles, providing that the chloride threshold for steel corrosion initiation is known. The parameters, including the rate coefficient $K_{sq}$ and the intercept *b*, can contain more fundamental information, such as the concrete intrinsic properties and interactions between concrete and environments. Once these fundamental factors are revealed, these two parameters are expected to provide more support both for concrete materials design and service life design for marine concrete structures.

### *4.3.2 Resistivity*

The electrical resistivity alone can be taken as indicator of durability. As an empirical indicator, the following Table 4.5 of equivalence with the apparent diffusion coefficient can be used in function of the exposure classes to chlorides.

This table has been deduced from the following equation where the $k_{\mathrm{Cl}}$ used for each exposure class can be found in Table 4.4 and in which the effective diffusion coefficient has been substituted by the apparent one,

$$\rho_{ef} = \frac{K_l}{D_a} \tag{4.44}$$

**Table 4.5** Equivalence between $D_a$ and resistivity $\rho$ when used as empirical indicator

| $D_a$ ($cm^2/s$) | $\rho$ ($\Omega \cdot m$) | | |
|---|---|---|---|
| | XSD1 | XSD2 | XSD3 |
| $1 \times 10^{-8}$ | 100 | 170 | 250 |
| $2 \times 10^{-8}$ | 50 | 85 | 125 |
| $3 \times 10^{-8}$ | 34 | 57 | 84 |
| $5 \times 10^{-8}$ | 20 | 34 | 50 |

If no reference value of the $D_a$ is available, the required resistivity for a given service life can be simplified by neglecting the reaction and aging factors in Eqs. (4.35) and (4.36), through

$$\rho_{ef} = \frac{t_l \cdot K_l}{x^2} \tag{4.45}$$

### 4.3.3 *Formation Factor/Tortuosity*

The concept of the formation factor ($FF$) was introduced by Archie [6] as part of his work on the electrical properties of porous rocks. Archie proposed a relationship known as Archie's law, which relates the electrical resistivity of a porous material to its porosity and the resistivity of the fluid filling pores. Due to the porous structure of concrete materials, a formation factor is also introduced for indicating the characteristics of the pore structure and evaluating the mass transport properties, such as chloride transport,

$$FF = \frac{\sigma_{ps}}{\sigma_p} = \frac{D_{ps}}{D_p} = \frac{1}{\varphi\beta} \tag{4.46}$$

where $\sigma_{ps}$ is the electrical conductivity of pore solution and $\sigma_p$ is the electrical conductivity of paste. $D_{ps}$ and $D_p$ are the diffusivity of the ions in the bulk pore solution and cement paste, respectively. As Eq. (4.46) shows, it is correlated to both porosity ($\varphi$) and pore connectivity ($\beta$) [28, 68], which also has a relation with ionic conductivity and the ion transport properties as highlighted in [69].

Conductivity of concrete can be measured by Wenner method. The electrical conductivity of pore solution is possible to be calculated based on ion concentrations in squeezed pore solution or by the chemical composition of binders and moisture content as described in Ref. [31]. According to Huang et al. [32], $FF$ is a practical parameter for onsite performance evaluation, and they found that it could be used to estimate the $D_{RCM}$ of pure cement pastes and the blended binder system based on the following expression,

$$D_{RCM} = \frac{D_{ps}}{FF} \tag{4.47}$$

Here $D_{ps}$ is the chloride migration coefficient in the bulk pore solution, often assumed to be $2.03 \times 10^{-9}$ ($m^2/s$) [59], but it could be smaller due to the interaction of chlorides with other ions in the pore solution. As Fig. 4.14 shows, the above equation works well in prediction of rapid chloride migration coefficient of both pastes and the blended mortars.

Pore space in cement-based materials is highly chaotic, with pore sizes that vary across a broad range. But what makes porous media particularly complex is that the paths for fluid flow and ions are not straight, but tortuous and meandering. A molecule often must traverse a path that is several times longer than the straight line between its original source and its destination. The concept of tortuosity was introduced by Carman [10]. The purpose was to match the permeability, calculated based on the bundle of capillary tubes, to experimental data [21]. In effect, all relevant length scales were divided by the tortuosity to make them longer than what was provided by the bundle of capillary tubes so as to approximate the actual length of the paths that fluid molecules travel through the pore space.

In the literature, tortuosity has been defined as either a geometric parameter or one that is related to the hydraulic, electrical, or diffusive properties [16, 21, 52, 54]. If we consider the ion transport in connection with the direct-current (DC) electrical conductivity. The electrical tortuosity, $\tau_e$, in Carman's definition, can be defined as [15, 66],

$$\tau_e = \left( \frac{\langle L_e \rangle}{L_s} \right)^2 \tag{4.48}$$

where $\langle L_e \rangle$ is the average path length for electrical flow, and $L_s$ is the straight-line length for electrical flow through the medium. The electrical tortuosity can be also inferred by measuring the electrical resistivity of a medium as the product of its porosity $\varphi$ and the formation factor [17, 21, 55]:

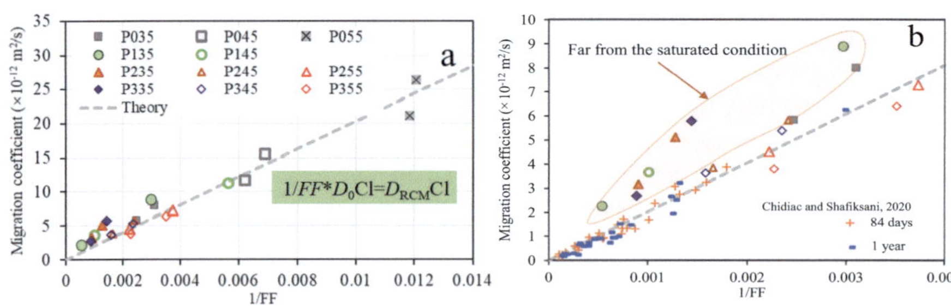


**Fig. 4.14** The correlation between the chloride migration coefficient and $1/FF$: a—data in this paper at the curing time of 28 and 90 days; b—data in [14] was used and adopted after some normalizations. Reproduced from [31] under the terms of the Creative Commons [CC BY 4.0], © [2024] [Elsevier]

$$\tau_e = \varphi \times FF \tag{4.49}$$

The tortuosity is a fundamental parameter in describing chloride transport in concrete, as it significantly influences the effective diffusivity and consequently the rate of ion migration through the concrete's porous network, reflecting the complexity of the pore structure. A higher value of tortuosity indicates a more convoluted pathway, thereby reducing the effective diffusivity of chlorides. Tortuosity has been used in models such as Eq. (4.31) to describe the properties of pore structure of cement-based materials.

### 4.3.4 Other Indicators

As aforementioned, the chloride transport in marine environment will be more or less influenced by the complex surface processes such as carbonation in ATM, drying wetting in SPL and leaching in TID/SUB (cf. Fig. 3.4), making the chloride profiles deviated from the pure diffusion curves, cf. Fig. 4.11. In these cases, forcing a fitting curve of pure diffusion, such as Fick's law, will distort the physical meaning of the regressed chloride diffusivity and surface chloride concentration. Here a new indicator, total chloride ingress quantity $M_{\mathrm{Cl}}$, is attempted to describe this global chloride ingress including the surface processes, cf. Fig. 4.15. This section is dedicated to the theoretical exploration and practical fitting on real chloride profiles from long-term exposure data.

Let us look at the pure diffusion of chlorides described by the analytical solution, from Fick's law with constant boundary condition, through Eq. (4.1). The chloride flow can be expressed as,

$$J = -D\frac{\partial C}{\partial x} = c_S\sqrt{\frac{D}{\pi t}}\exp\left(-\frac{x}{2\sqrt{Dt}}\right)^2 \tag{4.50}$$

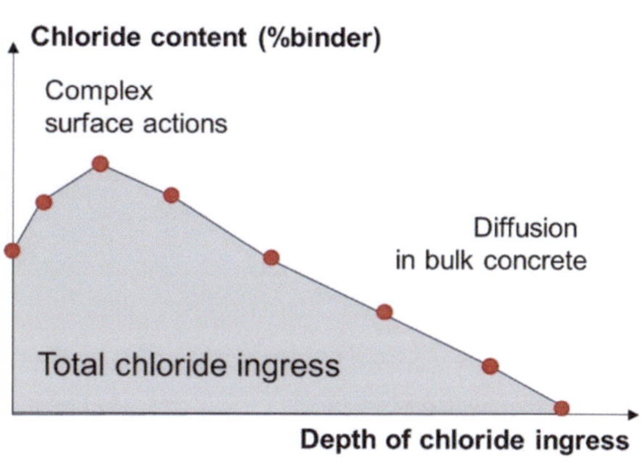


**Fig. 4.15** Illustration of the total chloride ingress for the chloride profiles consisting of a complex near-surface no-diffusion segment and diffusion segment in bulk concrete

And the chloride flow at the boundary $x = 0$ writes,

$$J|_{x=0} = \frac{c_s\sqrt{D}}{\sqrt{\pi t}} \tag{4.51}$$

The total amount of chlorides entering the specimen during the interval [0, t] can be obtained integrating chloride flow over time,

$$M_{Cl}(t) = \int_0^t Jdt = \frac{2c_s}{\sqrt{\pi}}\sqrt{Dt} \tag{4.52}$$

This is the theoretical expression of the total chloride ingress for a pure diffusion process, related to the surface concentration of chloride (constant) and the chloride diffusion coefficient (constant). Now the challenge is to use it as an engineering indicator for chloride profiles illustrated in Fig. 4.15. Take as an example the chloride profiles from exposure sites (splashing and tidal zones) of Cape Town [40]. Figure 4.16 gives the fitting results for splashing and tidal zones between the quantity $M_{Cl}/C_s$ and $\sqrt{Dt}$, knowing that the theoretical slope is 1.13 according to Eq. (4.52). Form the results, it can be seen that a linear correlation exits for both SPL and TID chloride profiles, the TID remains very close to pure diffusion, and the SPL profiles are deviated from pure diffusion due to the surface complex processes.

Another analysis using this quantity consists in determining the long-term chloride diffusivity using Eq. (4.52) using the known total chloride ingress $M_{Cl}$, surface chloride concentration $Cs$, and the profiling age $t$. On the exposure site Träslövsläge Harbor, Sweden, concrete specimens were measured for their chloride migration coefficient at 180 days. Thus, the correlation between the long-term diffusivity, deduced from Eq. (4.52) at different time interval, and the 180-day migration coefficient can be studied. The results are showed in Table 4.6, which shows better linear correlation in SUB but less linear in SPL.

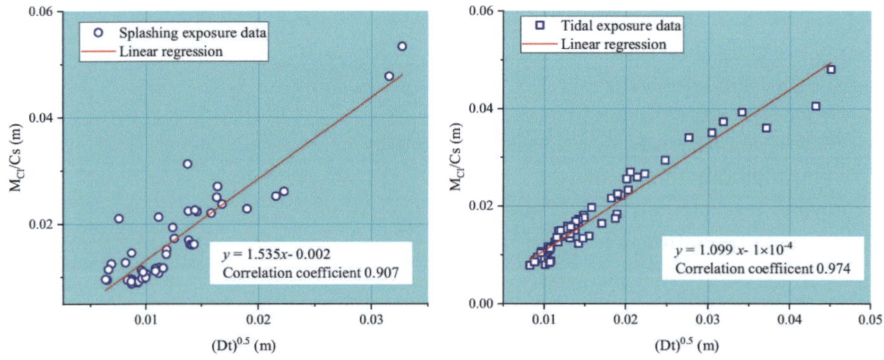


**Fig. 4.16** Correlation between $M_{Cl}/C_s$ and $\sqrt{Dt}$ for chloride profiles in splashing zone (left) and tidal zone (right) at the exposure site of Cape Town, South Africa

**Table 4.6** Linear regression between the predicted diffusion coefficient and 180d migration coefficient for concrete specimens in Träslövsläge Harbor site, Sweden

| Regression parameter | Submerge (SUB) zone | Splashing (SPL) zone |
|---|---|---|
| Slope | 0.384 | 0.25 |
| Intercept | 5.05 | 1.51 |
| *P* coefficient | 0.831 | 0.795 |
| *P*-value | $9.06 \times 10^{-29}$ | $6.24 \times 10^{-18}$ |

Form the above two small scale analyses of chloride profiles using the total chloride ingress as indicator, one can see that it bears meaningful results for engineering use and it avoids the distortion of regressed parameters. Certainly, more data are needed for further exploration of this indicator and its possible use in the interpretation of long-term exposure data in marine environments.

## 4.4 Chapter Summary

This chapter has provided a comprehensive review and assessment of various models developed for interpreting long-term field exposure data, particularly with respect to chloride ingress in concrete structures exposed to marine environments. We firstly provided a synthetic analysis of models based on Fick's 2nd law, followed by a discussion of more complex models that account for physical and chemical processes, including updated models in the works from RILEM TC 270-CIM and RILEM TC 289-DCM. These models represent an important tool for predicting chloride ingress and its implications for reinforcement corrosion. Benchmarking results were reviewed to provide a structured comparison of the models, offering insights into their accuracy, applicability, and limitations. These emphasize the importance of model validation against long-term field data to ensure their reliability in real-world applications.

Beyond traditional modelling approaches, this chapter introduced new indicators such as resistivity, the formation factor, tortuosity, and other new factors from chloride profiles, which strengthen the available modelling and provide practical metrics for engineering applications. These alternative indicators promote the use of simplified and valid methods to assess the durability of RC in real exposure environment. In conclusion, this chapter provided an overall view of both established and emerging models in interpretation and exploitation of field exposure data. The benchmarking of these models, combined with the introduction of practical indicators, serves as a step towards more reliable and sustainable service life predictions, while also identifying areas where further research and model development are necessary.

## References

1. Andrade, C.: Calculation of initiation and propagation periods of service life of reinforcements by using the electrical resistivity. In: International RILEM Symposium on Concrete Science and Engineering: A Tribute to Arnon Bentur, pp. 23–30. RILEM Publications SARL (2004)
2. Andrade, C.: Calculation of chloride diffusion coefficients in concrete from ionic migration measurements. Cem. Concr. Res. **23**, 724–742 (1993)
3. Andrade, C., Castellote, M., d'Andrea, R.: Measurement of ageing effect on chloride diffusion coefficients in cementitious matrices. J. Nucl. Mater. **412**, 209–216 (2011)
4. Andrade, C., Kropp, J.: Advances in model for RILEM TC-178 "testing and modelling chloride penetration in concrete". In: Third IABMAS Workshop on Life-Cycle Cost Analysis and Design of Civil Infrastructure Systems, International Association of Bridge Maintenance and Safety (IABMAS), 24–26 March 2003, Lausanne, pp. 123–132 (2003)
5. Andrade, C., Tavares, F., Castellote, M., Petre-Lazars, I., Climent, M.A., Vera, G.: Comparison of the chloride models: the importance of surface concentration. In: 2nd International Symposium on Advances in Concrete through Science and Engineering, 11–13 September 2006, Quebec City, Canada. RILEM Publications SARL (2006). https://doi.org/10.1617/2351580028.015
6. Archie, G.E.: The electrical resistivity log as an aid in determining some reservoir characteristics. Trans. AIME **146**, 54–62 (1942)
7. Bamforth, P.B.: Enhancing reinforced concrete durability Guidance on selecting measures for minimising the risk of corrosion of reinforcement in concrete. Technical report 61: part 1, Concrete Society, Blackwater, UK (2004)
8. Baroghel-Bouny, V., Wang, X., Thiery, M., Saillio, M., Barberon, F.: Prediction of chloride binding isotherms of cementitious materials by analytical model or numerical inverse analysis. Cem. Concr. Res. **42**, 1207–1224 (2012). https://doi.org/10.1016/j.cemconres.2012.05.008
9. Cai, R., Han, T., Liao, W., Huang, J., Li, D., Kumar, A., Ma, H.: Prediction of surface chloride concentration of marine concrete using ensemble machine learning. Cem. Concr. Res. **136**, 106164 (2020)
10. Carman, P.C.: Fluid flow through granular beds. Trans. Inst. Chem. Eng. Lond. **15**, 150–156 (1937)
11. Castellote, M., Andrade, C., Alonso, C.: Measurement of the steady and non-steady-state chloride diffusion coefficients in a migration test by means of monitoring the conductivity in the anolyte chamber. Comparison with natural diffusion tests. Cem. Concr. Res. **31**, 1411–1420 (2001)
12. Chatterji, S.: On the applicability of Fick's second law to chloride ion migration through Portland cement concrete. Cem. Concr. Res. **25**, 299–303 (1995)
13. Cherif, R., Hamami, A.E.A., Aït-Mokhtar, A., Bosschaerts, W.: Thermodynamic equilibria-based modelling of reactive chloride transport in blended cementitious materials. Cem. Concr. Res. **156**, 106770 (2022). https://doi.org/10.1016/j.cemconres.2022.106770
14. Chidiac, S.E., Shafikhani, M.: Electrical resistivity model for quantifying concrete chloride diffusion coefficient. Cem. Concr. Compos. **113**, 103707 (2020). https://doi.org/10.1016/j.cemconcomp.2020.103707
15. Childs, E.C.: An Introduction to the Physical Basis of Soil Water Phenomena. Wiley, New York (1969)
16. Clennell, M.B.: Tortuosity: a guide through the maze. Geol. Soc. Lond. Spec. Publ. **122**, 299–344 (1997)
17. Coleman, S.W., Vassilicos, J.C.: Transport properties of saturated and unsaturated porous fractal materials. Phys. Rev. Lett. **100**, 035504 (2008)
18. Collepardi, M., Marcialis, A., Turriziani, R.: Penetration of chloride ions into cement pastes and concretes. J. Am. Ceram. Soc. **55**, 534–535 (1972)
19. Dehwah, O.H.A., Xi, Y.: Theoretical model for the coupling effect of moisture transport on chloride penetration in concrete. Cem. Concr. Res. **177**, 107431 (2024). https://doi.org/10.1016/j.cemconres.2024.107431

20. Delgado, J., Silva, F.A.N., Azevedo, A.C., Silva, D.F., Campello, R.L.B., Santos, R.L.: Artificial neural networks to assess the useful life of reinforced concrete elements deteriorated by accelerated chloride tests. J. Build. Eng. **31**, 101445 (2020)
21. Dullien, F.A.: Porous Media: Fluid Transport and Pore Structure. Academic, San Diego, US (2012)
22. Engelund, S., Mohr, L., Edvardsen, C.: General guidelines for durability design and redesign: DuraCrete - probabilistic performance based durability design of concrete structures, Contract BRPR-CT95-0132, project BE95-1347, The Netherlands (2000)
23. FIB: Model code for service life design, fib Bulletin 34. Fédération Internationale du Béton, Lausanne (2006)
24. Fjendbo, S., Sørensen, H.E., De Weerdt, K., Geiker, M.R.: The square root method for chloride ingress prediction—applicability and limitations. Mater. Struct. **54**, 61 (2021). https://doi.org/10.1617/s11527-021-01643-8
25. Frederiksen, J.M., Geiker, M.: On an empirical model for estimation of chloride ingress into concrete. In: Andrade, C., Kropp, J. (eds.) Proceedings of the 2nd International RILEM Workshop "Testing and Modelling Chloride Ingress into Concrete", 11 September 2000, Paris France, pp. 355–371 (2000)
26. Frederiksen, J.M., Geiker, M.R.: Chloride ingress prediction: part 1: analytical model for time dependent diffusion coefficient and surface concentration. In: Concrete Modelling–CONMOD'08 Proceedings of the International RILEM Symposium, RILEM publications, 25–28 May 2008, pp. 257–282. Delft, The Netherlands (2008)
27. Frederiksen, J.M., Nilsson, L.O., Poulsen, E., Sandberg, P., Tang, L., Andersen, A.: HETEK, A system for estimation of chloride ingress into concrete, theoretical background. The Danish Road Directorate, Demark Ministry of Transport (1997)
28. Garboczi, E.J.: Permeability, diffusivity, and microstructural parameters: a critical review. Cem. Concr. Res. **20**, 591–601 (1990). https://doi.org/10.1016/0008-8846(90)90101-3
29. Helfferich, F.G.: Ion Exchange. Dover Publications, New York (1995)
30. Hosseinzadeh, M., Mousavi, S.S., Hosseinzadeh, A., Dehestani, M.: An efficient machine learning approach for predicting concrete chloride resistance using a comprehensive dataset. Sci. Rep. **13**, 15024 (2023)
31. Huang, L., Tang, L., Löfgren, I., Olsson, N., Yang, Z.: Real-time monitoring the electrical properties of pastes to map the hydration induced microstructure change in cement-based materials. Cem. Concr. Compos. **132**, 104639 (2022). https://doi.org/10.1016/j.cemconcomp.2022.104639
32. Huang, L., Tang, L., Löfgren, I., Olsson, N., Yang, Z., Li, Y.: Moisture and ion transport properties in blended pastes and their relation to the refined pore structure. Cem. Concr. Res. **161**, 106949 (2022). https://doi.org/10.1016/j.cemconres.2022.106949
33. Kari, O.-P., Elakneswaran, Y., Nawa, T., Puttonen, J.: A model for a long-term diffusion of multispecies in concrete based on ion–cement-hydrate interaction. J. Mater. Sci. **48**, 4243–4259 (2013)
34. Koenders, E., Imamoto, K., Soive, A.: Benchmarking Chloride Ingress Models on Real-Life Case Studies—Marine Submerged and Road Sprayed Concrete Structures: State-of-the-Art Report of the RILEM TC 270-CIM. Springer Nature, Switzerland (2022)
35. Li, K.F., Pang, X.Y., Dangla, P.: Thermo–hydro-ionic transport in sea immersed tube tunnel. Tunn. Undergr. Space Technol. **58**, 147–158 (2016)
36. Li, K.F., Zeng, J.J., Tang, L.P., Sørensen, H.E., Castro Borges, P., Geiker, M.R., Pedersen, M.T., Zhang, P., Surana, S., Maddalena, R., Wang, J., Andrade, C., Baroghel-Bouny, V., Martirena-Hernández, F., Geng, G., Kovler, K., Wang, S.: Long-term field exposure of structural concretes in marine environment: state-of-the-art review by RILEM TC 289-DCM. Mater. Struct. **55**, 205 (2022). https://doi.org/10.1617/s11527-022-02027-2
37. Li, K.F., Zhao, F.Y., Zhang, Y.M.: Influence of carbonation on the chloride ingress into concrete: theoretical analysis and application to durability design. Cem. Concr. Res. **123**, 105788 (2019). https://doi.org/10.1016/j.cemconres.2019.105788

38. Lin, C., Lee, L.: A two-ionic-parameter approach for ion activity coefficients of aqueous electrolyte solutions. Fluid Phase Equilib. **205**, 69–88 (2003)
39. Liu, Y., Cao, Y., Wang, L., Chen, Z.-S., Qin, Y.: Prediction of the durability of high-performance concrete using an integrated RF-LSSVM model. Constr. Build. Mater. **356**, 129232 (2022)
40. Mackechnie, J.R.: Predictions of reinforced concrete durability in the marine environment. Ph.D. thesis, University of Cape Town, South Africa (1995)
41. Mejlbro, L.: The complete solution of Fick's second law of diffusion with time-dependent diffusion coefficient and surface concentration. In: Sandberg, P. (ed.) Durability of Concrete in Saline Environment, Cementa, Sweden, pp. 127–158 (1996)
42. Meng, Z., Liu, Q., Ukrainczyk, N., Mu, S., Zhang, Y., De Schutter, G.: Numerical study on the chemical and electrochemical coupling mechanisms for concrete under combined chloride-sulfate attack. Cem. Concr. Res. **175**, 107368 (2024). https://doi.org/10.1016/j.cemconres.2023.107368
43. Nguyen, T.Q., Baroghel-Bouny, V., Dangla, P.: Prediction of chloride ingress into saturated concrete on the basis of a multi-species model by numerical calculations. Comput. Concr. **3**, 401–422 (2006)
44. Nilsson, L.-O.: Prediction models for chloride ingress and corrosion initiation in concrete structures. In: Nilsson, L.O. (ed.) P-01:6 Chalmers University of Technology, Sweden (2001)
45. NordTest: Concrete, mortar and cement-based repair materials: chloride migration coefficient from non-steady-state migration experiments (NT Build 492), Nordtest, Espoo, Finland (1999)
46. O'Neill Iqbal, P., Ishida, T.: Modeling of chloride transport coupled with enhanced moisture conductivity in concrete exposed to marine environment. Cem. Concr. Res. **39**, 329–339 (2009). https://doi.org/10.1016/j.cemconres.2009.01.001
47. Papadakis, V.G., Vayenas, C.G., Fardis, M.N.: Fundamental modeling and experimental investigation of concrete carbonation. ACI Mater. J. **88**, 363–373 (1991)
48. Pitzer, K.S.: Activity Coefficients in Electrolyte Solutions. CRC Press, Boca Raton, US (1991)
49. Poulsen, E.: 13 concrete diseases, how they develop, proceed and are prevented (in Danish), SBI Betong 4. The Danish Institute of Building Research, Hørsholm, Denmark (1985)
50. Pradelle, S., Thiéry, M., Baroghel-Bouny, V.: Comparison of existing chloride ingress models within concretes exposed to seawater. Mater. Struct. **49**, 4497–4516 (2016)
51. Saetta, A.V., Scotta, R.V., Vitaliani, R.V.: Analysis of chloride diffusion into partially saturated concrete. ACI Mater. J. **90**, 441–451 (1993)
52. Sahimi, M.: Flow phenomena in rocks: from continuum models to fractals, percolation, cellular automata, and simulated annealing. Rev. Mod. Phys. **65**, 1393 (1993)
53. Samson, E., Lemaire, G., Marchand, J., Beaudoin, J.J.: Modeling chemical activity effects in strong ionic solutions. Comput. Mater. Sci. **15**(3), 285–294 (1999)
54. Scheidegger, A.E.: The Physics of Flow Through Porous Media. University of Toronto Press, Toronto (1957)
55. Schopper, J.R.: A theoretical investigation on the formation factor/permeability/porosity relationship using a network model. Geophys. Prospect. **14**, 301–341 (1966)
56. Soive, A., Tran, V.Q., Baroghel-Bouny, V.: Requirements and possible simplifications for multi-ionic transport models – case of concrete subjected to wetting-drying cycles in marine environment. Constr. Build. Mater. **164**, 799–808 (2018). https://doi.org/10.1016/j.conbuildmat.2018.01.015
57. Tang, L.P.: Engineering expression of the ClinConc model for prediction of free and total chloride ingress in submerged marine concrete. Cem. Concr. Res. **38**, 1092–1097 (2008). https://doi.org/10.1016/j.cemconres.2008.03.008
58. Tang, L.P.: Chloride ingress in concrete exposed to marine environment-field data up to 10 years exposure (No. SP Report 2003:16). SP Swedish National Testing and Research Institute, Borås, Sweden (2003)
59. Tang, L.P., Gulikers, J.: On the mathematics of time-dependent apparent chloride diffusion coefficient in concrete. Cem. Concr. Res. **37**, 589–595 (2007). https://doi.org/10.1016/j.cemconres.2007.01.006

60. Tang, L.P., Nilsson, L.-O.: Chloride binding capacity and binding isotherms of OPC pastes and mortars. Cem. Concr. Res. **23**, 247–253 (1993). https://doi.org/10.1016/0008-8846(93)90089-R
61. Tang, L.P.: Chloride transport in concrete-measurement and prediction. PhD Thesis, Chalmers University of Technology, Sweden (1996)
62. Tang, L.P., Boubitsas, D., Huang, L.M.: Long-term performance of reinforced concrete under a de-icing road environment. Cem. Concr. Res. **164**, 107039 (2023)
63. Tang, L.P., Nilsson, L.-O.: Service life prediction for concrete structures under seawater by a numerical approach. In: Durability of Building Materials & Components 7, vol. 1, pp. 97–106. Routledge, Oxfordshire (2018)
64. Tang, L.P., Nilsson, L.-O., Basheer, P.M.: Resistance of Concrete to Chloride Ingress: Testing and Modelling. CRC Press, New York (2011)
65. Thomas, M.D.A., Bentz, D.: Life 365–Computer program for predicting the service life and life-cycle costs of reinforced concrete structures exposed to chlorides. In: User Manual (Version 1.0. 0), Presented at the Nordic Mini Seminar & fib TG, US (2001)
66. Tindall, J.A., Kunkel, J.R., Anderson, D.E.: Unsaturated Zone Hydrology for Scientists and Engineers. Prentice Hall, New Jersey, US (1999)
67. Tran, V.Q.: Machine learning approach for investigating chloride diffusion coefficient of concrete containing supplementary cementitious materials. Constr. Build. Mater. **328**, 127103 (2022)
68. Tumidajski, P.J., Schumacher, A.S., Perron, S., Gu, P., Beaudoin, J.J.: On the relationship between porosity and electrical resistivity in cementitious systems. Cem. Concr. Res. **26**, 539–544 (1996). https://doi.org/10.1016/0008-8846(96)00017-8
69. Wilson, W., Georget, F., Scrivener, K.: Unravelling chloride transport/microstructure relationships for blended-cement pastes with the mini-migration method. Cem. Concr. Res. **140**, 106264 (2021). https://doi.org/10.1016/j.cemconres.2020.106264
70. Xie, M.J., Dangla, P., Li, K.F.: Reactive transport modelling of concurrent chloride ingress and carbonation in concrete. Mater. Struct. **54**, 177 (2021). https://doi.org/10.1617/s11527-021-01769-9

# Chapter 5
# Long-Term Exposure: Guideline for Practice

**Kefei Li, Junjie Zeng, Shengnian Wang, Luping Tang, Liming Huang, Henrik Erndahl Sørensen, Carmen Andrade, Riccardo Maddalena, Guoqing Geng, Pedro Castro Borges, Gilberto Cidreira Keserle, José Fernando Martirena-Hernández, Fragkoulis Kanavaris, André Valente Monteiro, Konstantin Kovler, Peng Zhang, Tushar Bansal, Talakokula Visalakshi, and Roberto Torrent**

This chapter is taken from RILEM publication in Materials and Structures (2024) 57:44 with permission to reproduce by Springer.

K. Li (✉)
Civil Engineering Department, Tsinghua University, Beijing, China
e-mail: likefei@tsinghua.edu.cn

J. Zeng · S. Wang
CCCC Fourth Harbor Engineering Co., Ltd., Guangzhou, China

L. Tang · L. Huang
Department of Architecture and Civil Engineering, Chalmers University of Technology, Gothenburg, Sweden

H. E. Sørensen
Building and Construction, Danish Technological Institute, Taastrup, Denmark

C. Andrade
International Center for Numerical Methods in Engineering (CIMNE), Barcelona, Spain

R. Maddalena
School of Engineering, Cardiff University, Cardiff, UK

G. Geng
Department of Civil and Environmental Engineering, National University of Singapore, Singapore, Singapore

P. C. Borges
Centro de Investigación y de Estudios Avanzados del IPN, Unidad Mérida, Mérida, Yucatán, México



K. Li et al. (eds.), *Long-Term Performance of Structural Concretes in Marine Exposure Conditions*, RILEM State-of-the-Art Reports Volume,
https://doi.org/10.1007/978-3-032-06959-7_5

## 5.1 Aim and Scope

The long-term exposure practice of concrete materials in marine environments provides a unique source for durability mechanism investigation and durability model validation [38]. The scope of this guideline covers the main features of design and operation aspects of exposure sites of concrete materials in marine environments. The principles stated in this guideline can, with careful examination, be applied to the long-term exposure practice of other engineering materials and other natural environments. This guideline aims to optimize the design and the operation of exposure sites, and help to generate more added values from long-term exposure practice.

Typically, the exposure sites or stations can be divided into two categories: exposure sites for general use [20, 45, 49] and those for specific projects [34, 35, 48]. The former is established to support investigating the durability mechanisms and performance of concrete materials while the latter is conceived to verify the durability performance of concrete materials in given projects. Accordingly, the general exposure sites are usually chosen to represent the environmental conditions of a given region. The project-based sites are normally built into the projects to use the environmental conditions identical to projects for exposure. Both categories of exposure sites are covered by this guideline.

The practice of long-term exposure sites in marine environments begins with the design of exposure sites, involving the choice of sites, design of exposure facilities and exposed specimens, placing and protection of exposed specimens. Then, the monitoring of exposure sites consists of monitoring of environmental conditions

---

G. C. Keserle
Englobe Corp, Québec, Canada

J. F. Martirena-Hernández
Facultad de Construcciones, Universidad Central de las Villas, Santa Clara, Villa Clara, Cuba

F. Kanavaris
Arup Group Ltd, London, UK

A. V. Monteiro
National Laboratory for Civil Engineering, Lisbon, Portugal

K. Kovler
National Building Research Institute—Faculty of Civil and Environmental Engineering, Technion—Israel Institute of Technology, Haifa, Israel

P. Zhang
Qingdao University of Technology, Qingdao, China

T. Bansal
Sharda University, Noida, Uttar Pradesh, India

T. Visalakshi
Mahindra University, Hyderabad, Telangana, India

R. Torrent
Materials Advanced Services Ltd., Buenos Aires, Argentina

and the performance of exposed specimens. The environmental conditions refer to the meteorological parameters, such as the temperature and the relative humidity, and the local contact conditions of specimens, such as the chloride accumulation rate. The monitoring of durability performance pertains to the continuous measurement of target performance parameters of exposure specimens, and the associated sampling and test methods. Both monitoring operations need to set up protocols for data management. Finally, systematic inspection and maintenance schemes are needed for the exposure facilities to maintain the operation.

Following this logic line, this recommendation is organized as follows: the design of exposure sites is described in Sect. 5.2, the monitoring of environmental conditions is provided in Sect. 5.3, the monitoring of durability performance of exposed specimens is clarified in Sect. 5.4, the data management is covered in Sect. 5.5, and the maintenance of exposure facilities is specified in Sect. 5.6. This recommendation is expected to help the engineers and the structure owners to rationalize and optimize their design and operation of the long-term exposure sites. The researchers can also benefit from this recommendation for a clearer vision on the long-term exposure practice in view of better interpretation and exploitation of exposure data.

## 5.2 Design of Exposure Stations

### *5.2.1 Choice of Exposure Site*

The exposure sites (stations) in marine environments refer to sheltered or open-air areas, exposed in coastal areas, with infrastructure and facilities allowing for engineering materials, such as concrete and reinforced concrete, to be investigated for their performance under the environmental actions. The design of the exposure sites begins with the choice of site (siting). The siting should consider, but not be limited to, the environmental conditions, infrastructure for exposure, accessibility to transport and to power, safety, security and costs. In addition, the duration of exposure and the severity of the environment can be factors for siting.

#### 5.2.1.1 Environmental Conditions

The environmental conditions refer to the local meteorology (temperature, humidity, precipitation, wind direction and speed, atmospheric $CO_2$ concentration and air-borne chlorides), and the seawater hydrology (sea levels, wave heights, salinity, sediment flushing). The environmental conditions of the chosen site should meet the needs of long-term exposure study. For general use sites, the environmental conditions should be representative of the defined region. For project-based sites, they should represent the exposure conditions of the projects or the defined parts of project.

#### 5.2.1.2 Infrastructure for Exposure

The exposure site should be situated in locations where the exposure infrastructure can be easily and conveniently constructed. Reuse of existing structures and facilities is recommended. If construction works will be involved, the entailed environmental and ecological impact, if any, should be minimized.

#### 5.2.1.3 Accessibility

The chosen site should be accessible to transport (roadway or waterway), fresh water and electric power. Also, the site should be convenient for specimen placement and retrieval.

#### 5.2.1.4 Safety

Extreme climates and environmental actions such as typhoons and earthquakes should be considered during site choice. Normally, the site choice should avoid these extreme conditions to protect the exposed infrastructure and specimens. If these climates cannot be avoided, e.g. sites set up in open seas, special measures should be taken to assure the safety of personnel, infrastructure and specimens.

#### 5.2.1.5 Cost Considerations

The choice of exposure site should consider the entailed costs related to the construction of infrastructure, if any, the operation and the maintenance of the exposure site. It is a good practice to combine exposure sites with existing or planned constructions, or to reuse decommissioned structures and facilities for exposure.

### *5.2.2 Exposure Infrastructure and Facilities*

The exposure infrastructure can be newly built for exposure purpose, or reuse existing structures and facilities, cf. Fig. 5.1. The service life of these structures should meet the requirement of exposure study, normally not less than 50 years. In marine environments, anti-corrosion measures are recommended to maintain and extend the service life of these structures.

The exposure infrastructure should allow for the placement and retrieval of specimens, the works of on-site testing and the maintenance activities. The exposure infrastructure should also provide the required exposure conditions for the exposure study. The marine exposure conditions are classified as submerged, tidal, splash and

**Fig. 5.1** Austefjord field station supported by an existing quay, Norway (left) and HongKong-Zhuhai-Macau Bridge exposure site built into the project, China (right). The exposure sites may be built on some existing structures, such as wharves in service, or designed and built as new structure. In either case, the service life of the structures should cover the expected operation duration of the exposure site. *Source* Li et al. 2024, Fig. 1. Reproduced with permission of Springer

atmospheric zones [15]. The correct design of these exposure zones within the exposure infrastructure needs the long-term recording of the seawater hydrology, and, if any, the shielding effect of the exposure infrastructure against the seawater flow.

The selected space for the submerged zone should ensure the specimens to be kept immersed in seawater permanently. The space for tidal zone should subject the specimens to periodic seawater tidal actions, whilst the space for splash zone should put specimens to the drying-wetting actions of the waves. Finally, the space for atmospheric zone should expose the specimens to marine saline atmosphere, cf. Fig. 5.2. The atmospheric zone may be further designed as heavy/light exposure zones depending on deposition rates of airborne chlorides and the shelter conditions. The detailed definition of different zones is provided in Chap. 1.

In addition to the main exposure structures, the safety facilities, such as stairs, guardrails, fences, and fastening chains, need to be designed properly according to their functions. Indicating and warning signs should be installed for spaces of different exposure zones.

### *5.2.3 Exposed Specimens*

The design of specimens should meet the requirements of the long-term exposure study, considering their geometry, numbers, exposure conditions, raw materials, curing and manufacture, labelling, sampling frequency and testing.

The geometry of specimens is determined by the study purpose. Specimens of large size are recommended when continuous sampling or coring are needed for the

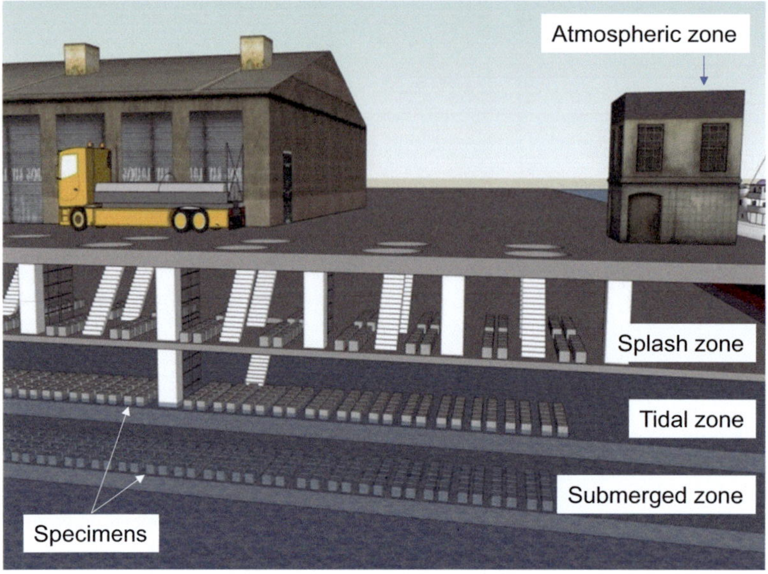


**Fig. 5.2** Illustration of the spaces designed for different exposure zones in Zhanjiang exposure site, China. The exposure site is situated in existing marine wharfs, using the existing structures to provide submerged, tidal, splash and atmospheric exposure conditions for concrete and reinforced concrete specimens. Due to the shielding effect of wharf structures, the local wave splashing is more relevant to this arrangement, and will be different from the splash actions in open sea. *Source* Li et al. 2024, Fig. 2. Reproduced with permission of Springer

exposure study, or when structural effects such as cracking also constitute part of the investigation. Specimens of small size can be used for single-property monitoring. For small-sized specimens, attention should be paid to the concrete homogeneity, and the minimum size of the specimen should be larger than 5 times the maximum size of aggregates used in concrete.

The number of specimens is determined upon the study purpose and the storage capacity of exposure site. Since large-size specimens can support multiple sampling, the required number of large-sized specimens can be less than that of small-sized specimens when continuous sampling or coring is needed. Duplicate specimens are recommended when the same parameter is to be checked for a same exposure age.

The raw materials and proportioning of concrete should conform to the requirements of the exposure study. Prior to transport to the exposure site, the concrete specimens should be subject to specific and well-defined procedures of curing. For general studies, standard curing to a maturity of 28d or 56d is recommended. If any specific curing condition is intentionally simulated, the concrete specimens should be cured under this specific scheme before exposure. It is recommended to characterize the concrete properties of fresh state and at standard curing ages in the laboratory.

When the specimens are ready for exposure, surface treatments are necessary for the study of ionic transport into concrete. All surfaces except the exposed one should be sealed. For specimens in submerged and tidal zones, epoxy-resin is recommended for sealing while, for splash and atmospheric zones, coatings with aging-resistance, such as acrylic- and silicone-based coatings, are recommended. For cubic specimens, the casting top and bottom surfaces are not recommended as the exposed surface.

The specimens placed in the exposure site should be labelled clearly and each specimen should bear a unique label. The label information should include the responsible institution, exposure zone, concrete mixture, casting date and study purpose. The label tag should be protected by sealing materials or made from resistant materials as the specimens are in direct contact with seawater. If necessary, the label tag can also include a QR code with more information available, cf. Fig. 5.3.

It is a good practice to establish a primitive mapping of the exposure site, containing the crude drawing of exposure site and the labelled positions of exposed specimens. Also, photographic documentation of specimens is recommended. Both measures help to minimize any possible loss of data due to missing of labels.

**Fig. 5.3** Large-sized specimens (0.2 m × 1 m × 2 m) from exposure site at Rødbyhavn, Denmark (left) and small-sized specimens (100 mm cubes) in Zhanjiang exposure site, China (upper right), and a label tag on exposed specimen (down right). *Source* Li et al. 2024, Fig. 3. Reproduced with permission of Springer

### *5.2.4 Placing and Protection of Exposed Specimens*

Exposed specimens from the same batch of concrete mixture or for the same study purpose, should be arranged and placed in the same unit of space (possibly covering different exposure zones) to facilitate their transport, placing and retrieving.

Necessary measures should be taken to protect the exposed specimens from extreme climates and vandalism. The fixation facilities, such as cages made from steel wires together and their anchorage, are recommended for small-sized specimens. These fixation facilities are better made with corrosion-resistant materials such as stainless steel, titanium alloy and galvanized steel. Special shielding and protecting measures should be taken if other disturbing factors arise, such as stray current or polluting agents in seawater.

## 5.3 Monitoring for Environmental Conditions

The aim of recording environmental conditions is to describe accurately the environmental envelope of the exposed specimens. The surface environment of concrete specimens includes also the interference effects of exposure infrastructure, even the concrete specimens themselves, in the local meteorology and hydrology. These conditions are crucial for characterizing the performance of concrete specimens in exposure and conducting supplementary experiments and modeling in laboratory [22, 52]. Further, if the exposure data are to be compared among different exposure sites, accurate recording of relevant environmental conditions becomes indispensable [39, 40]. In addition, extreme climate events should be recorded separately. The exposure zonation in marine environments is adopted as follows (altitude from low to high) [15]: the submerged zone (SUB), the tidal zone (TID), the splash zone (SPL), and the atmospheric zone (ATM). This zonation is detailed in Appendix A, and the involved durability processes can be referred to [38, 39].

### *5.3.1 Environmental Conditions*

The environmental conditions are represented by the following factors with their relevance, if any, to the exposure zones.

#### 5.3.1.1 Temperature

The temperature refers to both the atmosphere and the seawater enveloping the exposure specimens. Since both temperatures can have important temporal and spatial variations, it is recommended to monitor the temperature locally near the exposed specimens.

The temperature is related to exposure zones. In the SUB zone, the specimen temperature is equal to that of seawater; both the air and seawater temperatures are relevant to specimens exposed in TID and SPL zones; the temperature in ATM zone follows the air temperature. For accurate analysis of temperature, the thermal interaction between the environment and the concrete surface may be necessary, e.g. through an equivalent surface temperature [39].

#### 5.3.1.2 Sea (Water) Levels

The water levels refer to the high tide and low tide levels from the seawater hydrology recordings. Together with the wave height, the water levels help to determine the ranges for TID and SUB zones for shielded and unshielded conditions [18, 21]. Note that the water levels are not constant values but related to the return period of hydrology observation.

#### 5.3.1.3 Wave Height

The wave height refers to the distance between the crest and the trough of a wave, affected by local hydrology factors such as water depth. The wave height assumes also statistical properties from the recording of hydrology, and its value depends on the wave height-occurrence distribution. The wave height will largely determine the ranges of ATM and SPL. Note that in exposure sites with shielded conditions provided by the exposure infrastructure, the wave height can be largely different from the seawater hydrology, and its determination needs supplementary local recordings for the exposure infrastructure.

#### 5.3.1.4 Precipitation

Precipitation refers to the duration and amount of atmospheric precipitation from the meteorology recordings. Its value is highly correlated to the relative humidity in marine atmosphere. The precipitation is relevant to the humidity conditions for ATM zone and SPL zone.

#### 5.3.1.5 Time of Wetness (ToW)

The time of wetness pertains to the total time of the concrete surface wetting due to rain, surface condensation and water flowing from higher parts of the structure. This factor helps to quantify the effective wetting time after wetting events, a typical quantity for environment-material interaction [14]. A quantitative method has been proposed to evaluate ToW for concrete surface [51]. The factor is primarily relevant to ATM zone, and also relevant to SPL zone.

#### 5.3.1.6 Relative Humidity

The relative humidity depicts the relative vapor content in the ambient atmosphere, available from the local meteorology recordings. Note that for the exposure infrastructure with shielding effect, e.g. from atmosphere precipitation, the relative humidity may be different from the local meteorology recordings. This factor is relevant to ATM zone, and also to SPL zone because it scales to what extent the air is wet during the period without seawater contact.

#### 5.3.1.7 Wind Direction and Speed

The wind direction and speed refer to the wind characteristics from meteorology recordings. Again, for exposure infrastructure with shielding effects, the impact of wind speed and direction should be studied locally. The wind factors are relevant to ATM and SPL zones: the wind direction can determine the preferred directions of the airborne chloride deposition and the seawater splashing; the wind speed will determine the reach of airborne chlorides and the range of splashing action [44, 57].

#### 5.3.1.8 Salinity (Concentrations of Chlorides, Sulfates)

The salinity refers to the total amount of dissolved salts present in seawater, defined as the mass of salts, in grams, dissolved in 1 kg seawater and expressed in parts per thousand (‰). The sodium ($Na^+$) and chloride ($Cl^-$) ions are the predominant ionic species in seawater, followed by magnesium ($Mg^{2+}$) and sulfate ($SO_4{}^{2-}$) ions. The last three species bear aggressive nature for concrete materials. The concentrations of chlorides, sulfates and magnesium can be obtained through chemical analysis in laboratory from seawater samples. The salinity varies in terms of the water depth and coastal distance: higher salinity value is found in deep water and salinity becomes lower as one approaches the coast. Since the marine environment relates closely to seawater, the salinity of seawater is relevant for all exposure zones.

#### 5.3.1.9 $CO_2$ in Air

The $CO_2$ in air refers to the concentration of $CO_2$ present in the marine atmosphere, also available from the meteorology recordings. The $CO_2$ concentration value is usually noted in parts per million (ppm). This factor is important for the correct interpretation of chloride ingress in ATM and SPL zones because during the "dry" period the carbonation of surface layer will alter the chloride binding and transport in concrete [59].

#### 5.3.1.10 Air-Borne Chlorides

The airborne chlorides refer to the quantity of chlorides contained in the tiny seawater droplets sprayed into the atmosphere by the aerodynamics of marine air. This is the key source for the chlorides of ATM zone, and its quantity depends much on the wind direction and speed. Normally the measurement of airborne chlorides should be arranged close to the exposure site (specimens) and needs special experimental setups [3].

#### 5.3.1.11 Biological Growth

The biological growth refers to the marine colonizing organisms depositing on the concrete surface, usually by calcium carbonate formation [46]. The influence of biological growth on the durability of concrete is not yet well understood, but it seems to change the microclimate at the concrete surface, e.g. the growth of mussels and barnacles can coat the concrete surface and thus impede the chloride penetration. The biological growth occurs mainly in SUB zone, also possible in tidal and SPL zones.

### *5.3.2 Monitoring Methods*

The environmental conditions for exposure site are represented by the factors in the previous section. For a specific exposure site, these factors can either be obtained from local meteorological and hydrological recordings if these recordings represent the exposure site conditions, or from the monitoring on the exposure site. Table 5.1 summarizes the available monitoring methods, monitoring sensors and devices for these factors, together with the recommended data collection frequency. The relevance to the different exposure zones is also given in the table.

The monitoring methods in Table 5.1 are detailed in the following. The measurement of seawater salinity is based on the electrical conductivity measurement [50], deduced from an established relationship between the electrical conductivity and the salinity under given temperature and pressure. For the air-borne chlorides, the wet-candle method collects aerosols and particles retained on a wet wick, expressing the chloride deposition rate in mg/(day·m$^2$) [6]. A salt-laden mist sensor was developed to adapt the measurement in cold climate [31], the mortar disc method was used to simulate better the microclimate on concrete surface [25], and the capture tank method was conceived to collect the total load of air-borne chlorides [56]. The monitoring of biological growth can be performed through colorimetric measurements for

**Table 5.1** Monitoring methods for environmental factors in marine exposure sites

| Environmental factor | Method | Device/Sensor | Exposure zone | Recommended frequency |
|---|---|---|---|---|
| Temperature (air, seawater) | Sensor | Thermocouple<br>Resistance thermal detector<br>Thermistor<br>Semiconductor | All | 1/h |
| Sea level | Sensor | Pressure sensor<br>Ultrasonic sensor | SUB, TID | 1/h |
| Wave height | Sensor | Microwave radar<br>Laser sensor<br>MEMS sensor | SPL, ATM | 1/h |
| Precipitation | Sensor | Radar precipitation sensor<br>Laser weather sensor<br>Weighing precipitation gauge | SPL, ATM | 1/h |
| Time of Wetness (ToW) | Sensor | Resistivity sensor<br>WETCORR sensor<br>Leaf wetness sensor | SPL, ATM | 1/h |
| Relative humidity | Sensor | Capacitive sensor<br>Resistivity sensor | SPL, ATM | 1/h |
| Wind direction | Sensor | Wind vane sensor | SPL, ATM | 1/h |
| Wind speed | Sensor | Anemometer | SPL, ATM | 1/h |
| Salinity | Sensor | Conductivity sensor [50] | All | 1/day |
| Chloride/Sulfate | Device | Chemical analysis | SUB | 1/year |
| $CO_2$ concentration in air | Sensor | PAS sensor, NDIR sensor | SPL, ATM | 1/month |
| Air-borne chlorides | Device | Wet candle method [6], custom made device [25, 31, 56] | ATM | 1/month (first year), followed by 1/quarter |
| Biological growth | Visual | Colorimetric measurement [42], scraping and weighing [19] | SUB, TID, SPL | 1/year |

short-term growth [42], and scraping and weighing methods for long-term growth [19]. The living lives of all sensors should be determined before their installation, allowing for a timely replacement.

## 5.4 Monitoring of Exposed Specimens

### 5.4.1 Target Performance for Exposed Specimens

The performance of exposed specimens in marine environments refers to the resistance to different environmental actions, including chloride ingress, carbonation, freeze–thaw, sulfate attack, drying-wetting cycles, erosion, and the internal expansion reactions such as alkali-aggregate reaction and delayed ettringite formation. Some processes deteriorate the concrete materials while the others lead to the corrosion of embedded steel reinforcement. Since the processes deteriorating the concrete cover will also accelerate the steel corrosion, these processes are usually also regarded as the indirect causes for steel corrosion.

The resistance to chloride ingress is usually characterized by the chloride diffusivity of the concrete cover and the chloride ingress profiles. At given exposure ages, the chloride diffusivity scales how easily the chlorides can diffuse into concrete while the chloride ingress profile shows to what extent the chlorides have penetrated concrete. In practice, the chloride diffusivity has been found well correlated to the electrical resistivity (conductivity) [36], and thus electrical resistivity can be an alternative parameter for chloride diffusivity.

The resistance to concrete deteriorating processes is first characterized by the affected concrete properties such as strength, elastic modulus, and permeability. Other measurable parameters at macroscopic level include the volume change, the surface cracking and internal damage of specimens. For the drying-wetting and erosion actions, the chemical analysis of mineral composition in concrete cover is pertinent. The internal expansion reactions would need further petrography, microscopic, computerized tomography and chemical analyses.

The steel corrosion is characterized by the electrochemical stability of the embedded steel bars. The relevant parameters can include the half-cell potential and the polarization resistance of steel bars, and the electrical resistance of concrete cover. The carbonation changes the pore solution chemistry of concrete cover, and thus promotes the chloride ingress and the subsequent steel corrosion [37]. The parameter pertinent to carbonation is the carbonation depth into concrete cover. The distribution of mineral composition would also be pertinent to the carbonation resistance.

### 5.4.2 Sampling and Testing

Continuous sampling and testing are necessary to monitor the performance parameters of exposure specimens. A detailed study plan should be set up for the specimen sampling, retrieving and testing. Since most processes are diffusion-related, the observation frequency can scale to the square-root of exposure age. On the basis of 100a exposure duration, the sampling ages are recommended to follow 0.5a, 1a,

2a, 5a, 10a, 20a, 35a, 50a, 75a and 100a, and the sampling ages of 35a, 50a, 75a and 100a can be optional if the double number of specimens are available during the sampling ages between 0.5a and 20a.

#### 5.4.2.1 Sampling

The specimen sampling, by coring or cutting, should be conducted in a homogeneous way. The sampling operation should first record accurately the sampling position, the exposure condition (zone) of specimen and the specimen label. The number of samples should be enough to fulfil the required procedures for the target measurements, and avoid the local heterogeneity of concrete materials. The size of sampling surface depends on the study purpose and the drilling capacity as well. Special care should be taken when wet-saw coring or cutting is used in sampling for chloride profile measurement because the water jet in sampling operation can dissolve chlorides [9]. After a drilling or coring operation in large-sized specimens, the drilling holes should be re-filled and sealed.

#### 5.4.2.2 Testing

The obtained samples can either be tested on-site or transported to laboratory for testing. For laboratory testing, the samples should be kept in sealable plastic bags, protected from physical and chemical disturbance. Then, the tests on retrieved specimens and samples should be conducted in short delay to avoid physical or chemical alteration. The measurable parameters and the corresponding test methods are provided in Table 5.2.

The chloride diffusivity cannot be measured directly by laboratory methods such as rapid chloride permeability test (RCPT) [12], rapid migration test (RCM) [17], or Ponding methods [1, 47] because the exposed specimens contain already chlorides and these chlorides can affect the results. In this case, the rapid iodide migration (RIM) method may be suitable for testing ionic diffusivity in chloride contaminated specimens [32, 41]. The electrical resistivity of concrete is proposed as the alternative parameter to characterize the chloride diffusivity in Table 5.2.

Apart from the methods in Table 5.2, sensors can be placed in the specimens to monitor the steel corrosion using open circuit sensors, wireless sensors, or optic fibers [23, 27]. In marine environments, the sensors and corresponding equipment should be robust enough to survive the marine conditions and the possible contact with seawater. Moreover, these sensors should be sufficiently powered to be functional over long durations.

**Table 5.2** Measurable parameters and testing methods for the target performance

| Performance | Parameter | Testing method | Scope |
|---|---|---|---|
| Chloride ingress | Concrete resistivity (conductivity) | Bulk method [7, 33, 43] | Laboratory |
| | | Surface/Wenner probe [2] | In-situ |
| | Chloride profile | Grinding method, cutting and crushing method, dry drilling method [55] | Laboratory |
| | Ionic diffusivity | RIM [32, 41] | Laboratory |
| Carbonation | Carbonation depth | Phenolphthalein solution spray, or other reagents [16] | In-situ/laboratory |
| Steel corrosion | Half-cell potential | ASTM C876 [4] | In-situ/Laboratory |
| | Polarization resistance | ASTM C876 [4] | In-situ/Laboratory |
| | | GECOR [26], GalvaPulse [24], RapiCorr [53] | In situ, portable |
| | Concrete resistivity | See above | See above |
| Concrete property and damage | Compressive strength | ASTM C39 [11] | Laboratory |
| | Elastic modulus | ASTM C469 [13], UPV method [5] Fundamental frequency [8] | Laboratory |
| | Surface damage and cracking, discoloration | Visual, microscopy | In-situ/Laboratory |
| | Change in volume | (Vernier) callipers | In-situ/Laboratory |
| | Surface absorption rate | ASTM C1585 [10] | Laboratory |
| | Air permeability | Torrent device [54] | In-situ |
| | Moisture content | Tremax [30], embedded probes [28] | In-situ |

#### 5.4.2.3 Reporting

The reporting of exposure results may contain, but is not limited to, the information in Table 5.3. The data of recordings and measurements should be formatted in such a way that can be opened by commonly used software, such as text editors and spreadsheets. In addition, it is of great help to provide an authorized contact person of long-term availability, along with the information in Table 5.3.

**Table 5.3** Reporting of long-term exposure results

| Type | Item | Information | Nature |
|---|---|---|---|
| Environment | Exposure site | Station owner/operator (ID), exposure location (ID), opening year, status of station, exposure conditions provided, capacity of exposure station | Description |
| | Environmental factors | Height from the average level of seawater, and distance from the coastline | Description |
| | | Specified exposure zone (ATM, SPL, TID or SUB) | Description |
| | | Exposure duration (in years and months) | Digital |
| | | Average temperature and relative humidity of atmosphere surrounding the exposure specimens | Digital |
| | | Average temperature of seawater, and temperature of seawater at sampling | Digital |
| | | Salinity, or preferably chloride and sulfate concentrations, of the seawater | Digital |
| | | Other factors such as seawater chemistry, seawater hydrology, air-borne chlorides, CO2 concentration in air, precipitation, wind speed and direction, and biological growth at concrete surface | Digital/ Description |
| Materials | Concrete specimen label | Specimen ID, mix ID, specimen geometry, casting date, exposure date, exposure location/station (ID) | Digital/ Description |
| | Concrete mix proportioning and properties | Mix proportion, air content and chloride content of the fresh concrete, density and slump of fresh concrete, laboratory production and curing, compressive strength of concrete at 28d | Digital/ Description |
| | Raw materials and other properties | Type, chemical composition and physical properties of cement and mineral admixtures; type, maximum size and grade curve of aggregates; specimen surface cracking (if apply) | Digital/ Description |
| Performance | Basic properties (if apply) | Method for sampling; Size and geometry of sampled specimens; Method employed to assess the basic properties such as compressive strength and elastic modulus, and the obtained values | Digital/ Description |
| | Chloride ingress profiles | Date and Method for sampling; Size and geometry of sampled specimens; Method for analysis of chloride content, and Chloride content at any given sampling depth | Digital/ Description |

(continued)

**Table 5.3** (continued)

| Type | Item | Information | Nature |
|---|---|---|---|
| | Corrosion conditions of reinforcement | Type of reinforcement; Thickness of cover; Method for corrosion assessment, and Values and/or indexes from the measurements | Digital/ Description |
| | External/internal damages (if apply) | Date and Method for sampling; Size and geometry of sampled specimens; Method employed to assess the damage, and Values and/or indexes from the measurements | Digital/ Description |
| | Other properties (if apply) | Date and Method for sampling; Size and geometry of sampled specimens; Method employed to assess the properties, and Values and/or indexes from the measurements | Digital/ Description |

## 5.5 Data Management

The data management of long-term exposure is to collect, store, publish and reuse the data generated during the exposure practice. A successful data management can materialize the research efforts and generate added value. Some recent concepts for data management, such as FAIR data principles [58], provide a sound framework for data management. A data management system should consider the following elements: data and storage media, data transmission, data accessibility and sharing. Recommendations are given to these aspects on the basis of return of experiences from some major exposure sites.

### *5.5.1 Data Storage*

The data generated from the long-term exposure practice include the recordings on the environmental conditions, the information on concrete materials, and the testing results on the exposed specimens. The environmental data can be referred to the factors in Table 5.1, the information on concrete materials can be referred to the descriptions in Sect. 5.2.3, and the testing results can follow the description in Sect. 5.4.2. A standard presentation for exposure data can be referred to Tables 5.3 and 3.7.

The data storage modes include local storage and centralized storage. Local storage refers to the data collected directly from the exposure sites through sensors or recordings, or data and results from on-site or and off-site (laboratory) tests. Centralized storage refers to the systematic storage of the whole exposure datasets for a given exposure site. The data storage should be adapted to the change of storage technology. An electronic format is recommended for both modes of storage, and the data storage formats should be approved among the concerned parties. The data storages, local or centralized, should observe a pre-agreed protocol for data backup. In extreme cases where electronic format storage is not available, recordings and results in paper forms can be used, but a special guideline should be set up for the storage of these documents.

### 5.5.2 Data Transmission

The transmission of data is necessary when consolidating the data stored locally to the centralized storage. As the data transmission media are available and functional, the collected data, from on-site sensors or off-site laboratory tests, can be transmitted automatically as long as the sensors or the local storage media are connected to the centralized storage media through the same network. For in-site sensors, modulators are sometimes needed to convert the analogue signals to digital ones. If no transmission medium is available, the data transmission can only be done manually using movable storage media.

The transmission media for digital data can include wired transmission such as coaxial cables or optical fibre cables, and wireless transmission through microwaves (mobile phone) or radio waves (low power wide area networks, LPWAN). The cable transmission offers high transmission speeds, high security and reliability, but requires high power consumption and local access to a wired network. The mobile phone technology offers high transmission speeds, high security and reliability, but has high power consumption and requires local access to a mobile network. The LPWAN offers low power consumption, does not need access to wired or mobile network, but has a limited transmission speed and is not available in all countries [29].

### 5.5.3 Data Access and Sharing

For the centralized data storage, a systematic data management regulation should be established, clearly defining the data access and operation rights for the concerned persons. It is a good practice to carefully distribute different access rights of reading, copying, writing and deleting, especially to restrict the authorized persons with rights of writing and deleting. If some parts of data are confidential, they should be marked as such and only authorized persons should have access. For local data storage, the responsible institutions should apply a systematic data management and appoint authorized persons to record, maintain and transmit the local data. And the contact of these authorized persons is recommended to be included into the data reporting in Table 5.3.

A sharing policy should be agreed in advance among partners and investigators. The sharing of exposure data should comply with the active policies on intellectual property. A data sharing interface will be helpful with the main information of exposure data arranged into readable format.

## 5.6 Maintenance for Facilities of Exposure Site

### *5.6.1 General Management Requirements*

For an exposure site, specific logos are recommended to be set up in an easily visible location. The entrance into the exposure site should be approved by the operator, and the works in the exposure site should be registered and recorded. The works should be carried out according to the pre-determined plan which has been approved by the operator. No works should be performed during extreme climates.

The works in exposure site should respect the safety regulations. The personnel should be trained prior to the works according to relevant safety standards and maintaining operations. The safety equipment and items such as helmet, safety belt and life jacket should be worn properly. When electricity is used, special measures should be taken against the electricity leakage. The site should be recovered after works and all litters should be taken away and disposed. If chemicals are used during the works, ecological regulations should be followed.

### *5.6.2 Routine and Regular Inspection*

#### 5.6.2.1 Routine Inspection

Routine inspection is recommended to be performed on a yearly basis to check the appearance of the exposure facilities and specimens. It can be combined with the placing or retrieving of specimens. The routine inspection checks the following items and involves also the relevant maintenance if necessary.

The surface state of metal structures and facilities and their surface coatings, if any, need to be checked. Severely corroded parts should be replaced by new ones and broken coatings should be repaired. The specimens and their fixation facilities are to be checked. If the specimens are lost, new specimens should be prepared for substitution or the retrieval plan should be amended. If the fixation is failed, the facilities should be replaced.

The surface of exposure specimens is to be checked. In cases where the biological growth is deemed unacceptable, cleaning operations should be organized. In cases where the surface sealing and coating lose their integrity due to unexpected actions, repair works should be conducted immediately. For specimens with accessories such as loading devices and corrosion monitoring devices, these devices should be checked and any abnormality should be reported to the operator or the responsible institution.

When reinforced concretes are used as main exposure support, a monitoring system is recommended to be set up for these structures during their service life. Both video monitoring and photo monitoring are possible choices to transfer the images of the exposure site to remote terminals.

#### 5.6.2.2 Regular Inspection

Regular inspection is to check the integrity and the functionality of the exposure infrastructure and facilities. The frequency of regular inspection can be determined on the basis of the operation state of exposure site, normally no longer than 10 years. The regular inspection checks the following items and involves the relevant maintenance.

For reinforced concrete and steel structures to support the exposure practice, the regular inspection checks the structural integrity, loading resistance, structural displacement and deformation, and durability performance. If anti-corrosion protection and materials are used, they should be checked for their functionality. On the basis of the inspection results, a comprehensive assessment is to be conducted for the safety, serviceability and durability, and the corresponding maintenance works should be organized if necessary.

### *5.6.3 Special Inspection*

A special inspection of exposure sites should be carried out after special events such as ship collision, typhoon and earthquake. If the specimens are exposed in open air, they should be withdrawn before the passage of a hurricane to avoid their loss. The inspection should be done by a professional team. First, the access to power and water should be checked. Then, the safety of the main support structures should be assessed, and the relevant maintenance works should be performed if necessary. Lastly, the facilities and exposed specimens should be checked for their functionality and integrity. Any abnormality should be alerted to the operator and the responsible institution.

The special inspection occurs also for an exposure site approaching its design life or the end of exposure mission. The same items will be checked for the exposure site and the assessment should provide a basis for the continuation of the exposure site.

## References


1. AASHTO: Standard method of test for resistance of concrete to chloride ion penetration (T259-02), American Association of State Highway and Transportation Officials, Washington, DC, US (2002)
2. AASHTO: Standard method of test for surface resistivity indication of concrete's ability to resist chloride ion penetration (T358). American Association of State Highway and Transportation Officials, Washington DC, US (2019)
3. Alao, O.O.: Refinement of air-borne chloride exposure classes for RC structures in the Cape Peninsula. M.Sc. thesis, University of Cape Town (2015)
4. ASTM: Standard test method for half-cell potential measurements of reinforced concrete (C876). American Society for Testing and Materials, West Conshohocken, PA, USA (2015)
5. ASTM: Standard test method for pulse velocity through concrete (C597). American Society for Testing and Materials, West Conshohocken, PA, USA (2016)
6. ASTM: Standard test method for determining atmospheric chloride deposition rate by wet candle method (G140-02). ASTM International, West Conshohocken, PA, USA (2019)
7. ASTM: Standard test method for bulk electrical resistivity or bulk conductivity of concrete (C1876-19). ASTM International, West Conshohocken, PA, USA (2019)
8. ASTM: Standard test method for fundamental transverse, longitudinal, and torsional resonant frequencies of concrete specimens (C215). American Society for Testing and Materials, West Conshohocken, PA, USA (2019)
9. ASTM: Test method for obtaining and testing drilled cores and sawed beams of concrete (C42/C42M-20). ASTM International, West Conshohocken, PA, USA (2020)
10. ASTM: Standard test method for measurement of rate of absorption of water by hydraulic-cement concretes (C1585-20). ASTM International, West Conshohocken, USA (2020)
11. ASTM: Standard test method for compressive strength of cylindrical concrete specimens (C39/C39M-21). ASTM International, West Conshohocken, PA, USA (2021)
12. ASTM: Standard test method for electrical indication of concrete's ability to resist chloride ion penetration (C1202–22). ASTM International, West Conshohocken, PA, USA (2022)
13. ASTM: Standard test method for static modulus of elasticity and Poisson's ratio of concrete in compression (C469/C469M-22). ASTM International, West Conshohocken, PA, USA (2022)
14. Castro Borges, P., Véleva, L.P.: Time of wetness and temperature as tools to evaluate corrosion risk in concrete blocks exposed to a humid tropical environment. In: Dean, S.W., Hernandez-Duque De lgadillo, G., Bushman J.B. (eds.) Marine Corrosion in Tropical Environments, ASTM STP1399. ASTM International, West Conshohocken, US (2000)
15. CEB: Durable concrete structures – CEB design guide, Bulletin No. 183, Comité Euro-International du Béton (CEB), Lausanne, p. 120 (1992)
16. CEN: Testing hardened concrete. Part 12: determination of the carbonation resistance of concrete – accelerated carbonation method (EN 12390-12), European Standard, Brussels (2020)
17. CEN: Testing hardened concrete - Part 18: determination of the chloride migration coefficient (EN 12390-18:2021), European Committee for Standardization, Brussel, Belgium (2021)
18. China Industrial Standard: Corrosion prevention technical specifications for concrete structures of marine harbors engineering (JTJ 275–2000). China Transportation Publishing, Beijing (2000)
19. Coolen, J.W.P., Bittner, O., Driessen, F.M.F., van Dongen, U., Siahaya, M.S., Groot, W., Mavraki, N., Bolam, S.G., van der Weide, B.: Ecological implications of removing a concrete gas platform in the North Sea. J. Sea Res. **166**, 101968 (2020)
20. De Weerdt, K., Geiker, M., Orsakova, D.: Investigation of concrete from Solsvik field station. Traffic Safety, Environment and Technology Department, NPRA Reports, Norway (2015)
21. DNV: Design of offshore steel structures, general (LRFD method), offshore standard DNV-OS-C101, Det Norske Veritas, Høvik, Norway (2011)
22. Dousti, A., Shekarchi, M.: Effect of exposure temperature on chloride-binding capacity of cementing materials. Mag. Concr. Res. **67**(15), 821–832 (2015)

23. Duffó, G., Farina, S.: Development of an embeddable sensor to monitor the corrosion process of new and existing reinforced concrete structures. Constr. Build. Mater. **23**(8), 2746–2751 (2009)
24. Elsener, B., Klinghoffer, O., Frolund, T., Rislund, E., Schiegg, Y, Böhni, H.: Assessment of reinforcement corrosion by means of galvanostatic pulse technique. In: Blankvoll, A. (ed.) Proceedings of International Conference on Repair of Concrete Structures - From Theory to Practice in a Marine Environment, 28–30 May 1997, Svolvær, Norway, pp. 391–400 (1997)
25. Fagerlund, G.: Long-term mapping of the chloride load around a Swedish highway exposed to de-icing salt – with reference to reinforcement corrosion in concrete, Report TVBM-3127. LTH, Lund University, Division of Building Materials (2005)
26. Feliú, S., González, J.A., Feliú, S., Jr., Andrade, C.: Confinement of the electrical signal or in-situ measurement of polarization resistance in reinforced concrete. ACI Mater. J. **87**(5), 457–460 (1990)
27. Figueira, R.: Electrochemical sensors for monitoring the corrosion conditions of reinforced concrete structures: a review. Appl. Sci. **7**(11), 1157 (2017)
28. Huang, L.M., Tang, L.P., Löfgren, I., Olsson, N., Yang, Z.: Real-time monitoring the electrical properties of pastes to map the hydration induced microstructure change in cement-based materials. Cem. Concr. Compos. **132**, 104639 (2022)
29. Ioannou, K., Karampatzakis, D., Amanatidis, P., Aggelopoulos, V., Karmiris, I.: Low-cost automatic weather stations in the Internet of Things. Information **12**(4), 146 (2021)
30. Jacobs, F., Leemann, A., Denarie, E., Teruzzi, T.: Recommendations for the quality control of concrete with air permeability measurements. DETEC, Switzerland (2009)
31. Keserle, G.C., Sanchez, T., Conciatori, D., Chouinard, L.: Monitoring environmental and climatic exposure conditions for structures in cold regions. J. Cold Reg. Eng. **35**(2), 04021007 (2021)
32. Lay, S., Liebl, S., Hilbig, H., Schießl, P.: New method to measure the rapid chloride migration coefficient of chloride-contaminated concrete. Cem. Concr. Res. **34**(3), 421–427 (2004)
33. Layssi, H., Ghods, P., Alizadeh, A., Salehi, M.: Electrical resistivity of concrete. Concr. Int. **2015**, 41–46 (2015)
34. Li, K.F., Wang, P.P., Li, Q.W., Fan, Z.H.: Durability assessment of concrete structures in HZM sea link project for service life of 120 years. Mater. Struct. **49**, 3785–3800 (2016)
35. Li, K.F., Zhang, D.D., Li, Q.W., Fan, Z.H.: Durability for concrete structures in marine environments of HZM project: design, assessment and beyond. Cem. Concr. Res. **115**, 545–558 (2019)
36. Li, K.F., Qin, M.F., Gui, Q.: Durability properties of structural concretes containing secondary cementitious materials. Green Mater. **7**(1), 40–51 (2019)
37. Li, K.F., Zhao, F.Y., Zhang, Y.M.: Influence of carbonation on the chloride ingress into concrete: theoretical analysis and application to durability design. Cem. Concr. Res. **123**, 105788 (2019)
38. Li, K.F., Zeng, J.J., Tang, L.P., Sorensen, H.E., Castro-Borges, P., et al.: Long-term field exposure of structural concretes in marine environment: state-of-the-art review by RILEM TC 289-DCM. Mater. Struct. **55**, 205 (2022)
39. Lindvall, A.: Environmental actions on concrete exposed in marine and road environments and its response-consequences for the initiation of chloride induced reinforcement corrosion. Ph.D. thesis, Chalmers University of Technology (2003)
40. Lindvall, A.: Chloride ingress data from field and laboratory exposure – influence of salinity and temperature. Cem. Concr. Compos. **29**(2), 88–93 (2007)
41. Liu, J., Liao, C., Jin, H., Jiang, Z., Xie, R., Tang, L.P.: Electrically driven ionic transport in the RCM and RIM: investigations based on experiments and numerical simulations. Constr. Build. Mater. **331**, 127331 (2022)
42. Maury-Ramirez, A., De Muynck, W., Stevens, R., Demeestere, K., De Belie, N.: Titanium dioxide based strategies to prevent algal fouling on cementitious materials. Cem. Concr. Compos. **36**, 93–100 (2013)
43. McCarter, W.J., Chrisp, T.M., Starrs, G., Blewett, J.: Characterization and monitoring of cement-based systems using intrinsic electrical property measurements. Cem. Concr. Res. **33**(2), 197–206 (2003)

44. Meira, G.R., Andrade, M.C., Padaratz, I.J., Alonso, M.C., Borba, J.C., Jr.: Measurements and modelling of marine salt transportation and deposition in a tropical region in Brazil. Atmos. Environ. **40**(29), 5596–5607 (2006)
45. Moffatt, E., Thomas, M.: Performance of 25-year-old silica fume and fly ash lightweight concrete blocks in a harsh marine environment. Cem. Concr. Res. **113**, 65–73 (2018)
46. Natanzi, A.S., Thompson, B.J., Brooks, P.R., Crowe, T.P., McNally, C.: Influence of concrete properties on the initial biological colonization of marine artificial structures. Ecol. Eng. **159**, 106104 (2021)
47. Nordtest: Concrete hardened: accelerated chloride penetration (NT build 443), Nordtest method, Nordtest, Espoo, Finland (1995)
48. Poulsen, S.L., Sorensen, H.E., Jonsson, U.: Chloride ingress in concrete blocks at the Bodbyhavn marine exposure site – status after 5 years. In: 4th International Conference on Service Life Design for Infrastructures (SLD4), 27–30 August 2018, Delft, The Netherlands, pp. 192–203 (2018)
49. Roshore, E.: Cement durability program, long-term field exposure of concrete columns. US Army Engineer Waterways Experiment Station, Vicksburg, US (1972)
50. Souza Dias, F.: Salinity sensors (2020). http://www.coastalwiki.org/wiki/Salinity_sensors. Accessed 20 Oct 2022
51. Svennerstedt, B.: Surface moisture on façade materials. National Swedish Institute for Building Research, TN:16, Gävle, Sweden, p. 232 (1989). (in Swedish)
52. Tang, L.P.: Concentration dependence of diffusion and migration of chloride ions part 2. Experimental evaluations. Cem. Concr. Res. **29**(9), 1469–1474 (1999)
53. Tang, L.P., Fu, Y.: A rapid technique using handheld instrument for mapping corrosion of steel in reinforced concrete. Restor. Build. Monum. **12**(5–6), 387–400 (2006)
54. Torrent, R.J.: A two-chamber vacuum cell for measuring the coefficient of permeability to air of the concrete cover on site. Mater. Struct. **25**, 358–365 (1992)
55. Vennesland, Ø., Climent, M.A., Andrade, C.: Recommendation of RILEM TC 178-TMC: testing and modelling chloride penetration in concrete. Mater. Struct. **46**, 337–344 (2013)
56. Wattanapornprom, R., Ishida, T.: Modeling of chloride penetration into concrete under airborne chloride environmental conditions combined with washout effects. J. Adv. Concr. Technol. **15**, 126–142 (2017)
57. Wattanapornprom, R., Ishida, T.: Comprehensive numerical system for predicting airborne chloride generation and its ingress in concrete under actual environmental conditions. J. Adv. Concr. Technol. **16**(1), 18–35 (2018)
58. Wilkinson, M.D., Dumontier, M., Aalbersberg, I.J., Appleton, G., Axton, M., et al.: The FAIR guiding principles for scientific data management and stewardship. Scientific data **3**, 160018 (2016)
59. Xie, M.J., Dangla, P., Li, K.F.: Reactive transport modelling of concurrent chloride ingress and carbonation in concrete. Mater. Struct. **54**, 177 (2021)

# Chapter 6
# Long-Term Exposure: Support for Management

**Riccardo Maddalena, Kefei Li, Quanwang Li, Henrik Erndahl Sørensen, José Fernando Martirena, Alejandro Fernandez, Tingyu Hao, Carmen Andrade, Pedro Castro Borges, Luping Tang, Liming Huang, Junjie Zeng, Haicheng Yang, Mark Alexander, Saarthak Surana, and Peng Zhang**

R. Maddalena (✉)
School of Engineering, Cardiff University, Cardiff, UK
e-mail: MaddalenaR@cardiff.ac.uk

K. Li · Q. Li
Civil Engineering Department, Tsinghua University, Beijing, China

H. E. Sørensen
Building and Construction, Danish Technological Institute, Taastrup, Denmark

J. F. Martirena · A. Fernandez
Facultad de Construcciones, Universidad Central de las Villas, Santa Clara, Villa Clara, Cuba

T. Hao
Central Research Institute of Building and Construction, MCC, Beijing, China

C. Andrade
International Center for Numerical Methods in Engineering (CIMNE), Barcelona, Spain

P. C. Borges
Centro de Investigación y de Estudios Avanzados del IPN, Unidad Mérida, Mérida, Yucatán, México

L. Tang · L. Huang
Department of Architecture and Civil Engineering, Chalmers University of Technology, Gothenburg, Sweden

J. Zeng · H. Yang
CCCC Fourth Harbor Engineering Co. Ltd., Guangzhou, China

M. Alexander · S. Surana
Department of Civil Engineering, University of Cape Town, Cape Town, South Africa



K. Li et al. (eds.), *Long-Term Performance of Structural Concretes in Marine Exposure Conditions*, RILEM State-of-the-Art Reports Volume,
https://doi.org/10.1007/978-3-032-06959-7_6

## 6.1 Hong-Kong-Zhuhai-Macau (HZM) Sea Link Project

### *6.1.1 Background*

The Hong-Kong-Zhuhai-Macau (HZM) sea link project is one of the major infrastructures for the largest Chinese regional economic body, the Guangdong-Hong Kong-Macau bay area [28, 31]. The project is situated in the water region of Ling'ding Ocean on the southwestern coast of China mainland, connecting three major cities surrounding the bay area, Hong Kong, Zhuhai and Macau. The project consists of sea bridges of 28.8 km, two artificial islands in sea and one segment of immersed tube tunnel of 6.8 km [32]. The structures in this project are exposed to a rather aggressive marine environment and the design working life for the whole project is 120 years [27]. The design phase was finished in 2010 with the construction phase taking place from 2011 to 2018, when the whole infrastructure was open to traffic, as shown in Fig. 6.1. One of the major technical challenges is to assure the service life of 120 years for the whole structure, subjected to an aggressive marine environment with various deleterious species [29]. Concrete elements are widely used in the structures in this project with some typical elements presented in Table 6.1. To achieve the design working life of 120 years, a comprehensive durability design of such concrete elements was performed at both material and structural level, using model-based approaches to quantify chloride ingress and carbonation [33]. The durability design has been further reinforced through the durability assessment of concrete elements in construction with the help of data collected from in-site inspections and tests [30].

The concrete used in the HZM project has been designed to have high resistance to the chloride ingress in marine environment, i.e. low chloride diffusivity, and avoid such internal reaction as the alkali-aggregate or the delayed ettringite reaction [29]. The mixture design of these structural concretes should satisfy the target strength class, the workability, the durability requirement, and other early-age properties as well. Among these controlling factors, the durability requirements are represented by the chloride diffusivity measured by the rapid chloride penetration test (RCPT) method [39] obtained from the model-based durability design approach [29]. The mix proportions of the main structural concrete elements are presented in Table 6.2 with the chloride diffusivity measured in laboratory at given ages of 28 and 56 d. The low water to cement (w/c) ratio, around 0.3, and the ternary blend of Portland cement, fly ash and slag are the main features of these structural concretes.

After the construction phase, the project entered its service stage, expected to extend to 120 years. Although elaborated approaches have been adopted for the durability design with sufficient safety margins for the performance against the chloride ingress [30, 32], there still exists uncertainty for durability failure due to the nature of properties of concrete materials and external environmental factors. In

P. Zhang
Qingdao University of Technology, Qingdao, China

**Fig. 6.1** Hong Kong-Zhuhai-Macau sea link project situated on the south-eastern coast of China mainland across the Ling'ding Ocean and exposed to harsh marine environments. Image from public domain. *Source* https://www.zhxz.gov.cn/wlxz/tsjd/content/post_3590043.html (October 23, 2023)

**Table 6.1** Main reinforced concrete elements in concrete structures in the HZM project (RC stands for reinforced concrete, PC represents pre-stressed concrete)

| Structure | Element | Material | Construction |
|---|---|---|---|
| Cable-stayed bridges (navigable spans) | Integral Pylon | RC (base part) | Cast-in-place |
| | Box girder | RC (deck) | Cast-in-place |
| | Pier (auxiliary) | RC | Cast in place |
| | Bearing platform | RC | Cast in place |
| Steel box-girder bridges (non-navigable) | Box girder | PC (approach) | Cast-in-place |
| | Pier | RC | Prefabrication |
| | Bearing platform | RC | Prefabrication |
| Immerged tube tunnel | Tube segment | RC (intermediate) | Prefabricated |
| | | RC (approach) | Cast-in-place |
| | Anti-shock pier | RC | Prefabricated |
| Artificial islands | Retaining wall | Concrete | Cast-in-place |
| | Break water | Concrete | Prefabrication |

other words, one time design cannot fully eliminate uncertainties for the durability failure during a long service life of 120 years, and a systematic planning should be set up for the durability inspection/monitoring to collect data on real performance of concrete elements and structures. This allows a comprehensive durability assessment to master and quantify durability risk levels, and in turn define maintenance plans. Only through durability planning based on design, inspection and data on real performance and maintenance plan, a 120-year service life can be safely ensured [10]. The

**Table 6.2** Mixture design in laboratory for main structural concretes for concrete structures in HZM project [50]

| Requirement/Proportioning | | Bearing platform | Pier | Pylon | Box girder | Immersed tube tunnel |
|---|---|---|---|---|---|---|
| Requirement (design) | Strength class | C45 | C50 | C50 | C55 | C45 |
| | Cl diffusivity 28 d ($10^{-12}$ $m^2/s$) | ≤6.5 | ≤6.0 | ≤6.0 | ≤6.0 | ≤6.5 |
| | Cl diffusivity 56 d ($10^{-12}$ $m^2/s$) | ≤4.5 | ≤4.0 | ≤4.0 | ≤4.0 | ≤4.5 |
| Concrete mix proportions | Binder ($kg/m^3$) | 410 | 450 | 470 | 460 | 420 |
| | Water to cement ratio (-) | 0.34 | 0.31 | 0.32 | 0.31 | 0.35 |
| | Cement (% binder) | 35 | 40 | 45 | 45 | 45 |
| | Fly ash (% binder) | 25 | 25 | 20 | 20 | 25 |
| | Slag (% binder) | 40 | 35 | 35 | 35 | 30 |
| | Sand ratio (%) | 39 | 40 | 42 | 43 | 43 |
| Requirement (achieved) | Strength 28 d (MPa) | 58.1 | 67.5 | 69.2 | 70.3 | 59.9 |
| | Cl diffusivity 28 d ($10^{-12}$ $m^2/s$) | 4.6 | 4.5 | 5.0 | 4.4 | 4.7 |
| | Cl diffusivity 56 d ($10^{-12}$ $m^2/s$) | 2.9 | 2.9 | 2.8 | 2.5 | 2.5 |

inspection and monitoring of durability involve long-term exposure practices, along with pre-installed sensors for monitoring chloride ingress and steel corrosion. Such practice is in place for the HZM bridge and further presented.

### 6.1.2 Marine Exposure Site

The exposure site was designed and constructed on the western artificial island of the project during 2013, as shown in Fig. 6.2. The main function of this site is to accommodate the long-term exposure observation of the construction materials used in the project, including concrete and reinforced concrete specimens. The structure of this exposure site adopted a reinforced concrete frame with the design working life of 120 years, same as the project. The ultimate objectives of this exposure site and the relevant exposure practice are to deepen the knowledge of chloride ingress, calibrate scientific models and support the rational for the durability maintenance plan.

Four exposure zones were arranged. The atmospheric zone, positioned at > + 6.20 m, has exposure surface of 20 $m^2$; the splashing zone is situated at + 2.5 m and has a surface of 80 $m^2$; the tidal zone is arranged at + 0.00 m with an exposure surface of 30 $m^2$; the submerged zone is placed at −2.10 m with a functional surface

**Fig. 6.2** Exposure site constructed on the western artificial island of Hong Kong-Zhuhai-Macau Sea link project: global view (left) and top view (right). Courtesy of Mr. Jianbo Xiong

of 20 $m^2$. The site is designed to accommodate up to 2,000 specimens. The first group of specimens were fabricated along with the construction of concrete structures in the HZM project during November 2012–October 2013, and the exposure materials correspond to the structural materials used in the construction, including concrete, reinforced concrete and steel specimens. The typical exposure specimens and the corresponding structural elements are given in Table 6.3 with expected retrieval intervals. The concrete specimens are to be retrieved for laboratory test for chloride ingress profiles. The relevant experimental methods and procedure can be referred to Chap. 5.

During the years between 2020 and 2022, the exposure site was upgraded to retrofit the concrete structure of the exposure sites which were damaged by several significant typhoons, install sensors for environmental factors such as the detector for the disposition rate of environmental chlorides, and monitoring system for the corrosion of steel specimens.

### *6.1.3 Support for Long-Term Durability Planning*

The long-term durability planning of concrete structures in the HZM project for a 120-year service life design consists of (1) durability monitoring through the data collection from the exposure tests, the regular inspections and the durability-related sensors; (2) durability assessment based on these collected data; (3) the maintenance scheme driven by the durability assessment. These three aspects are interactive and driven by the durability monitoring, as shown in Fig. 6.3. Following this line, a comprehensive durability planning system has been developed for concrete structures in the HZM project using model-based approach to solve the durability assessment from the collected data.

So far, the data on the chloride ingress are available for 1, 3, 5 and 7 years, and the exploitation of these data will support the durability assessment. More specifically,

**Table 6.3** Main exposure specimens for concrete structures in HZM project

| Category | Study purpose | Target elements | Exposure zones | Retrieval scheme |
|---|---|---|---|---|
| Mechanical specimens | Long-term change of mechanical properties | Bearing platform, pier, pylon, immersed tunnel, box girder | ATM, SPL, TID, SUB | 1a, 3a, 5a, 10a, 30a, 50a |
| Chloride ingress | Long-term process of chloride ingress | Bearing platform, pier, pylon, immersed tunnel, box girder | ATM, SPL, TID, SUB | 1a, 2a, 3a, 5a, 10a, 20a, 30a, 50a, 70a, 90a, 120a |
| Loaded specimens | Coupled actions of chloride ingress and loading | Box girder | SPL, TID | 1a, 2a, 3a, 5a, 10a, 15a, 20a |
| Reinforced concrete specimens | Corrosion process of embedded steels | Bearing platform, pier, pylon, immersed tunnel, box girder | ATM, SPL, TID, SUB | Adapted to electrochemical measurements |
| Specimens with protections | Long-term effect of the different protections (stainless steel bars, epoxy-coated rebars, silane impregnation) | Bearing platform, pier, pylon, box girder | ATM, SPL, TID, SUB | Adapted to protection measures |

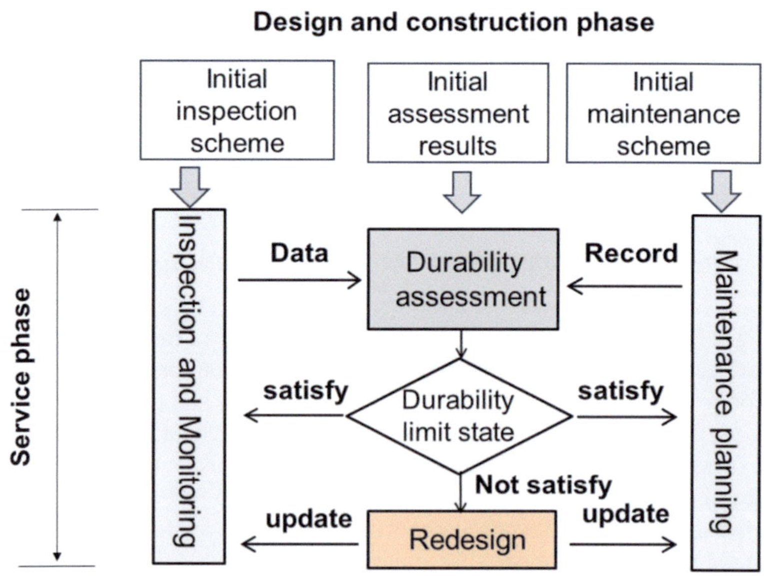


**Fig. 6.3** Interactive durability planning process involving durability inspection and monitoring, durability assessment and maintenance planning

these chloride profiles can be used to update the chloride ingress parameters and their statistical properties. Figure 6.4 (left) illustrates the obtained chloride profiles at 3 years for concrete specimens in splashing zone, corresponding to the structural concrete C50 for the prefabricated piers. Using appropriate updating schemes, the statistical properties of the parameters, surface chloride concentration and the chloride diffusivity, are updated and the corresponding failure probability is evaluated and shown in Fig. 6.4 (right). The results indicate that the surface chloride concentration and the chloride ingress rate, inferred from the exposure data, are both lower than the design values. Accordingly, the final failure probability for the reinforced concrete piers exposed in splashing zones becomes lower than the durability value. In other terms, the durability risk for corrosion (initiation) becomes lower through the updating from the exposure data.

Together with other collected data from durability monitoring, the durability assessment has been performed using the model-based probabilistic approach. The models used in the durability assessment can be referred to [30]. Then the maintenance planning is performed by optimizing the life cycle cost of the maintenance works [51]. Three basic elements are necessary: the durability model, the maintenance cost and the maintenance techniques. First, the durability model adopts the chloride ingress model in assessment [30], uses the corrosion initiation as the durability limit state (DLS) and adopts 1% as the maintenance limit for failure probability. The maintenance cost includes the cost for maintenance operation and the indirect cost, if any, to users, and converts the future costs to present time through the social discount rate [13]. The maintenance techniques adopted in the durability planning for concrete elements include the surface silane impregnation [43] and the concrete surface chloride extraction through electrochemical method [12], with their corresponding direct and indirect costs considered. The durability planning was then performed through the durability assessment and the choice of the maintenance schemes, maintenance techniques and their apply interval during service life, through optimising the total life cycle cost.

The durability assessment, considering the collected durability data, gives failure probability far lower than the expected maintenance limit of 1%, as shown in Table 6.4. Thus, these concrete elements can be considered as maintenance-free during their service lives of 120 years. However, considering the maintenance scheme will be continuously updated by the durability monitoring and potential local defects of concrete surface, a basic durability maintenance scheme is still proposed as follow: the silane impregnation is to be (re-)applied on the surface of concrete elements in atmospheric, splashing and tidal zones at the service ages of 60 and 90 years.

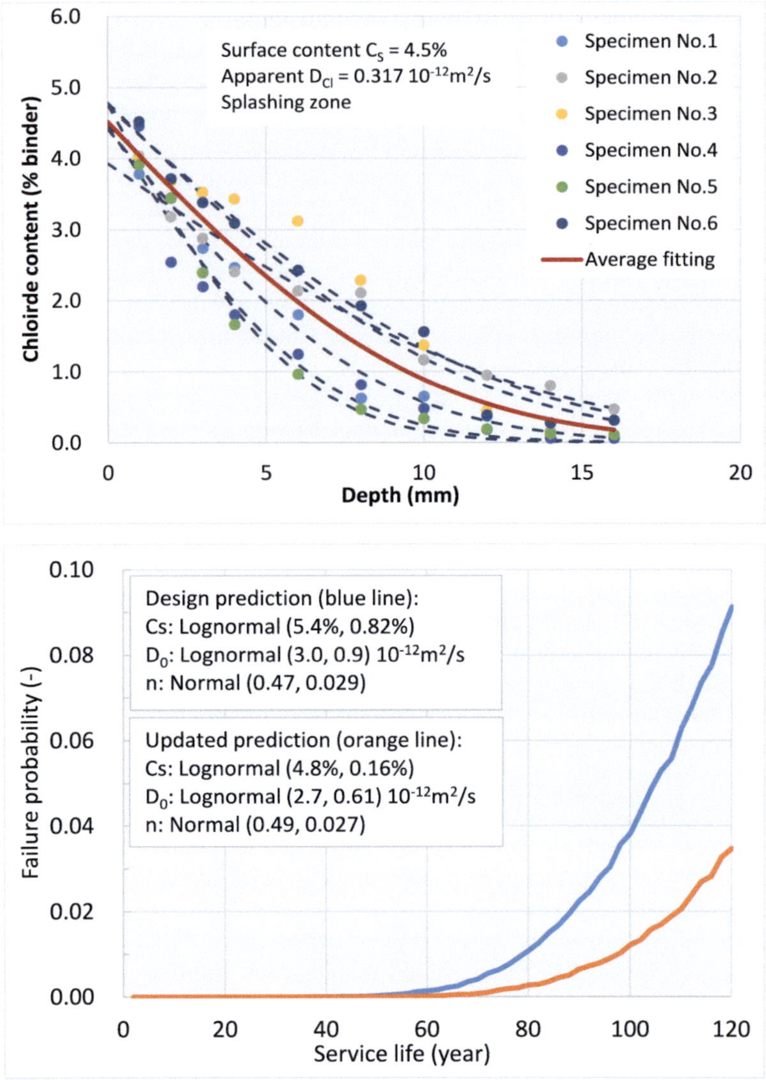


**Fig. 6.4** Regression of exposure data for chloride ingress after 3 years of exposure (up) and updating of durability assessment for RC piers in non-navigable bridges (down) for exposure in splashing zones. The updating scheme considers the change of surface chloride content Cs, the initial chloride diffusion coefficient $D_0$ and the ageing factor of diffusivity n. *Source* Li et al. [28, 31], Fig. 7. Reproduced with permission of Springer

**Table 6.4** Failure probability of typical concrete elements for corrosion initiation at 120 years

| Structure | Element | Exposure | Protection | Failure probability (%) | Reliability index (−) |
|---|---|---|---|---|---|
| Navigable span (cable-stayed bridge) | Pylon (upper, middle parts) | External/ATM | Silane | <0.01 | >4.0 |
| | | Internal/ATM | / | 0.035 | 3.39 |
| | Pylon (lower part) | External/ATM | SS, SI | 0 | >5.0 |
| | | Internal/ATM | / | 0.037 | 3.37 |
| | Trans-beam | ATM | SI, SS | 0 | >5.0 |
| | Pier (>+8.0 m) | External/ATM | SI, ECR | <0.01 | >4.0 |
| | | Internal/ATM | / | 0.052 | 3.28 |
| | Bearing platform | External/SPL | SS, SI | 0 | >5.0 |
| | | External/SUB | SS, SI | 0 | >5.0 |
| Non-navigable bridges | Pier (>+8.0 m) | External/ATM | SI, ECR | 0 | >5.0 |
| | | Internal/ATM | / | 0 | >5.0 |
| | Pier (<+8.0 m) | Internal/ATM | SI | 0 | >5.0 |
| | | External/SPL | ECR, SI | 0.108 | 3.07 |
| | | External/SUB | ECR, SI | 0 | >5.0 |

## 6.2 Coastal Infrastructure in Denmark

### *6.2.1 Background*

Denmark's extensive coastline and maritime climates create an ideal environment for studying concrete durability in marine environments. Positioned between the North Sea and the Baltic Sea, Denmark's coastal infrastructure faces challenges such as varying salinity, temperature fluctuations, and changing wind conditions. To address these, Denmark has established field exposure sites and conducted research on existing structures. This case study explores how long-term data from field exposure sites and old bridges can be used to develop and calibrate service life models and thus improve the durability and reliability of marine infrastructures. Particularly, the Danish exposure sites focus on assessing chloride ingress, reinforcement corrosion, and overall concrete durability under field conditions. Denmark hosts several concrete field exposure stations, including Rødbyhavn and Hirtshals (marine) and Hveens Boulevard (road). This section present data from the Rødbyhavn site, available at the Expert Centre for Infrastructure Materials (https://www.expertcentre.dk), and three old marine bridges, the Vejlefjord bridge, the Alssund bridge and the Farø bridges [40]. These cases help to gain insights into long-term concrete durability and capture complex material-environment interactions that laboratory experiments cannot replicate.

## 6.2.2 *Marine Exposure Sites*

### 6.2.2.1 Rødbyhavn Site

The Rødbyhavn marine exposure site, situated at 54.6548° N and 11.3529° E, was established in 2010 by Femern A/S to support the Fehmarnbelt Fixed Link project between Denmark and Germany. The seawater has an average chloride concentration of approximately 7 g/L. Water temperatures range from 0 to 22 °C, averaging at 9 °C. The site includes 15 large concrete blocks with varied mix designs exposed to seawater, cf. Figs. 1.7 and 5.3 (left), and 6 instrumented blocks for real-time corrosion monitoring. Each block measures 2 m in height, 1 m in width and 0.2 m in depth.

The mix designs incorporate three different cements, including low alkali sulphate resistant Portland cement (CEM I 42.5 N—SR5 EA), rapid hardening ordinary Portland cement (CEM I 52.5 N) and blast furnace slag cement (CEM III/B 42.5 N), and three different supplementary cementitious materials, including fly ash, silica fume and ground granulated blast furnace slag (GGBS). Chloride ingress measurements have been conducted after 0.5, 2, 5, and 10 years of exposure. The chloride profiles were analysed by extracting cores from both SUB and SPL zones of the blocks, followed by profile grinding and chloride content analysis using potentiometric titration [41].

### 6.2.2.2 Old Bridges

The Vejlefjord bridge, located at approximately 55.7045° N latitude and 9.5691° E longitude, is a key structure in Denmark, spanning the Vejle Fjord, as shown in Fig. 6.5. Completed in 1980, this concrete box girder bridge has endured over 40 years, with significant marine exposure to SPL and TID zones. Slag cement was utilised in the SUB, SPL and TID zones, with the concrete mix design detailed in Table 6.5. Chloride profiles have been measured and analysed on Ø100mm concrete cores sampled from SUB zone after 17, and 34 years of exposure.

The Alssund bridge in Fig. 6.6, located at approximately 54.9270° N latitude and 9.9077° E longitude, connects the island of Als with the mainland of Denmark over the Alssund strait. This bridge, completed in 1981, is made of long-span concrete beams exposed to marine conditions. The concrete mix design is shown in Table 6.5. Chloride profiles have been measured and analysed on Ø100 mm concrete cores sampled from SUB zone after 31 years of exposure.

The Farø bridges, comprising two bridges connecting the islands of Zealand and Falster in Denmark, as shown in Fig. 6.7, are located at approximately 54.9792° N latitude and 11.9875° E longitude. These bridges, completed in 1985, include a high bridge and a low bridge, both exposed to marine conditions. The concrete mix design is shown in Table 6.5. Chloride profiles have been measured and analysed on Ø100 mm concrete cores taken from the submerged zone after 11, 22 and 30 years of exposure.

**Fig. 6.5** Vejlefjord bridge in Denmark. Image reproduced with permission of the Road Directorate, Denmark. *Source* Courtesy of Lene Højris Jensen

**Table 6.5** Concrete mix proportioning and concrete cover of SUB zone for three marine bridges

| Characteristics | Vejlefjord bridge | Alssund bridge | Farø bridges |
|---|---|---|---|
| Cement (kg/m$^3$) | 400 | 335 | 370 |
| Fly ash (kg/m$^3$) | 0 | 0 | 45 |
| Silica fume (kg/m$^3$) | 0 | 0 | 0 |
| Fine aggregate (kg/m$^3$) | 590 | 741 | 560 |
| Coarse aggregate (kg/m$^3$) | 1135 | 1129 | 1205 |
| Water (kg/m$^3$) | 180 | 145 | 157 |
| Air (estimated) (%) | 5 | 5 | 5 |
| Water to binder (w/b) ratio | 0.45 | 0.43 | 0.38 |
| Nominal concrete cover (mm) | 50 | 50 | 50 |

## 6.2.3 *Durability Testing to Support Maintenance*

This section describes the methods used to measure and interpret chloride penetration in concrete based on analysis of concrete cores taken from Rødbyhavn sites. Chloride profiles were measured by grinding material layers parallel to the exposed surfaces of the cores. Chloride content was determined through potentiometric titration, and profiles were corrected for paste content variations using calcium profiles. Chloride transport parameters, such as the chloride diffusion coefficient ($D_a$), the surface

**Fig. 6.6** Alssund bridge in Denmark. Image reproduced with permission of the Road Directorate, Denmark. *Source* Courtesy of Lene Højris Jensen

**Fig. 6.7** Farø dual bridges in Denmark (the high bridge). Image reproduced with permission of the Road Directorate, Denmark. *Source* Courtesy of Lene Højris Jensen

chloride concentration ($C_s$), and the penetration parameter ($K_{Cr}$), were determined by fitting through the Fick's 2nd law of diffusion, details see Chap. 4. The determination of chloride threshold values is essential for predicting the onset of reinforcement corrosion in concrete structures. These values represent the critical concentration of chlorides surrounding the reinforcement at which corrosion initiates.

Three instrumented concrete blocks with different binders were designed for detailed monitoring of chloride ingress and corrosion. Each block is reinforced with 8 vertical steel bars with concrete covers to the steel bars being 20, 30, 40, 50, 60, 70, 80, and 90 mm. Four anode ladders are placed in each block, two in SPL zone and two in SUB zone, with discrete anodes at depths of 10, 20, 30 and 40 mm. Each anode connects to a data system that monitors electrical half-cell potential (HCP) and linear polarisation resistance (LPR), providing real-time corrosion data [45]. The blocks are partially submerged in seawater, allowing for chloride ingress studies and their impact on embedded reinforcement. Chloride threshold values at Rødbyhavn are determined through (1) continuous monitoring of the real-time corrosion data, (2) periodic sampling and measuring of chloride profiles; (3) correlating the corrosion data from anodes to the chloride profiles, and (4) determining the threshold chloride concentration for corrosion initiation at the first significant corrosion activity.

Corrosion monitoring at Rødbyhavn indicates real-time HCP measurements identify corrosion initiation effectively. Combining these observations with chloride profile measurements allows determination of site-specific chloride threshold values. Corrosion initiation detected on anodes with small covers confirms the capacity of these techniques to provide valuable input for service life modelling. Experiences from Rødbyhavn suggest incorporating anode ladders in new structures to determine local chloride threshold values under service conditions can significantly improve service life predictions. Ongoing research at Rødbyhavn underscores the importance of accurate measurement techniques.

### *6.2.4 Service Life Models to Support Maintenance*

Service life models are essential tools in predicting the long-term performance and durability of concrete structures in marine environments. Various types of service life models based on prediction of chloride ingress, their underlying principles, and their applications in real-world scenarios are discussed in Chap. 4. This section reviews the Femern model and the square root method, and their application to the selected infrastructures.

#### 6.2.4.1 Femern Model

The Femern model, developed specifically for the Fehmarnbelt Fixed Link project, incorporates a comprehensive approach to predicting chloride ingress by considering multiple factors for chloride transport in concrete [26]. The Femern model originates from the Mejlbro-Poulsen Model, developed in the Danish HETEK-project [25]. The basis for the Femern model is the "complete solution" of Fick's 2nd Law [24, 38], which utilises the ψ-functions to replace the error function in the traditional solution with the detailed description of model given in Sect. 4.1.1.4. Here only the main equations and parameters are recalled in the following equation. The chloride profile is expressed as:

$$C(x,t) = C_i + \left(S_p \cdot \tau^p\right) \cdot \psi_p\left(\frac{x}{2\sqrt{\tau \cdot t_{ex} \cdot D_{av,ex}}}\right) \tag{6.1}$$

where $C(x, t)$ is the chloride concentration at depth ($x$) and concrete age ($t$); $S_p$ is a parameter for time-dependent surface chloride concentration (boundary condition); $p$ is an exponent mainly depending on environment and binder type [$0 \leq p \leq 1$]; $t_{ex}$ is the concrete age at the beginning of exposure (time from casting to first exposure); $D_{av,ex}$ is the achieved transport coefficient characterising a chloride profile after natural exposure starting at $t_{ex}$; $\tau$ is a time parameter which can be expressed as,

$$\tau = \left(\frac{t}{t_{ex}}\right)^{1-\alpha} - \left(\frac{t_{ex}}{t}\right)^{\alpha} \tag{6.2}$$

where $\alpha$ is a parameter describing the decrease with time of the achieved chloride diffusivity.

The expression in Eq. (6.1) can be used to model chloride profiles when the four decisive parameters $\alpha$, $D_{av,ex}$, $S_p$ and $p$ have been determined. The approach undertaken by [26] is to start with estimation of chloride diffusion coefficients and surface concentrations after 1 year and 100 years exposure ($D_1$, $D_{100}$, $C_1$ and $C_{100}$) and then use these parameters to calculate $\alpha$, $D_{av,ex}$, $S_p$ and $p$. This comprehensive model accounts for the non-linear nature of chloride ingress and provides a more accurate prediction for long-term exposure scenarios. It incorporates the effects of varying environmental conditions and concrete properties, making it particularly useful for large infrastructure projects where precise service life predictions are critical.

The Femern model was initially calibrated by [26] on chloride profiles from Träslövsläge Field Exposure site in Sweden with exposure age ranging from 0.6 to 10.3 years [47] and then recalibrated with new chloride profiles from the same exposure site with exposure age up to 20.8 years [11]. In 2021–2023, a major update was undertaken by [46] and the model was recalibrated once more on new added chloride profiles from the Rødbyhavn exposure site with exposure age up to 10 years. So far, the model is calibrated on 153 chloride profiles from Träslövsläge site and 160 chloride profiles from Rødbyhavn site, covering 84 profiles in ATM zone, 72 profiles in SPL zone and 157 profiles in SUB zone.

Chloride profiles measured on existing structures have been used to evaluate the Femern model's ability to predict chloride profiles by inputting the concrete mix design, the exposure zone, and the exposure time. The predicted chloride profiles and model parameters are given in Fig. 6.8 and Table 6.6, using the concrete mix data in Table 6.5, for the three bridge cases: Vejle Fjord Bridge, Alssund Bridge, and the Farø Bridges.

In all three cases the measured chloride profiles are performed on concrete cores extracted from the submerged zone. The chloride profiles measured on the two bridges made with Portland cement concretes showed good agreement with the model predictions. The Femern model accurately predicted the shape and depth of chloride penetration in the inner part of chloride profiles for different concrete mixes. The prediction for the bridge made with slag cement concrete underestimates the chloride ingress to some extent, which may be caused by a limited number of chloride profiles available of the model calibration.

#### 6.2.4.2 Square Root Method

The square root method has been reviewed in Sect. 4.1.3.1, offering a simplified yet effective means of estimating chloride penetration [40]. The fundamental equation of the square root method is,

$$x(t, C_r) = a_{Cr} \cdot \sqrt{t - t_{ex}} + b_{Cr} \tag{6.3}$$

where $x(t, C_r)$ is the depth of chloride penetration at time $(t)$ for a reference concentration $(C_r)$; $a_{Cr}$ is a constant ingress rate representing the rate at which chlorides penetrate the concrete; $b_{Cr}$ is a constant accounting for an initial different ingress rate than described by $a_{Cr}$.

The square root method was applied to the bridge case (Farø bridges) and exposure specimens in Fig. 6.9 with a high correlation degree (average $R^2$ of 0.96) between the penetration depth of a reference chloride concentration and the square root of exposure time. It is thus confirmed that this method is particularly suitable for long-term predictions where the diffusion process dominates. Recent advancements in machine learning have introduced new methods for predicting chloride ingress, as reviewed in Sect. 4.1.3.5. These models use large datasets from field and laboratory studies to train algorithms that can predict chloride profiles with high accuracy, one such model can be found in [22].

#### 6.2.4.3 Limitation of Models

The Femern model and the square root method, along with the models reviewed in Chap. 4, have their strengths and limitations. For precise and accurate predictions, it is essential to consider the specific context and the potential limitations. The Femern model offers a more detailed approach by considering the non-linear nature of chloride ingress but requires more comprehensive data for calibration for a given exposure situation. The square root method offers a practical and straightforward approach to predicting chloride ingress in concrete, with strong empirical support

**Fig. 6.8** Prediction of chloride profiles by Femern model compared to measured chloride profiles for **a** Vejlefjord bridge, **b** Alssund bridge and **c** Farø bridge

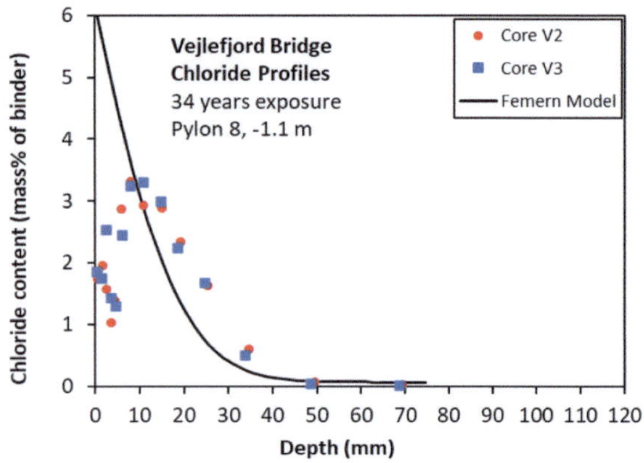


(a)

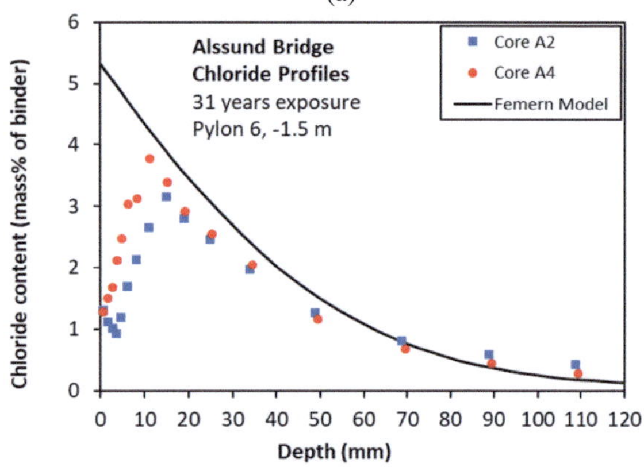


(b)

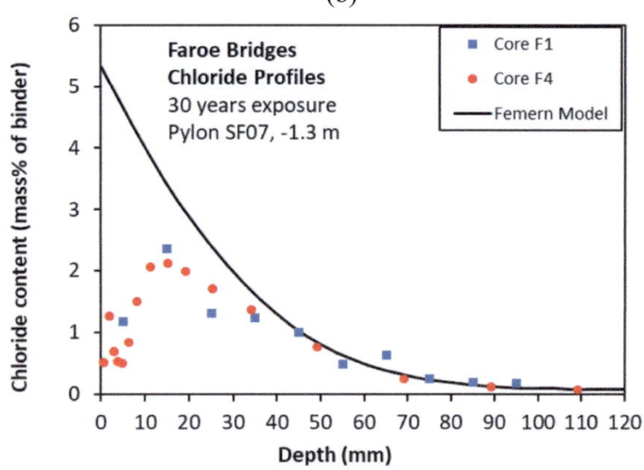


(c)

**Table 6.6** Femern model parameters for bridge cases

| Bridge | $\alpha$ (−) | $D_{av,ex}$ (mm²/year) | $S_p$ (mass % binder) | $p$ (−) |
|---|---|---|---|---|
| Vejlefjord bridge | 0.329 | 29 | 3.27 | 0.155 |
| Alssund bridge | 0.329 | 280 | 2.84 | 0.155 |
| Farø bridges | 0.329 | 153 | 2.86 | 0.155 |

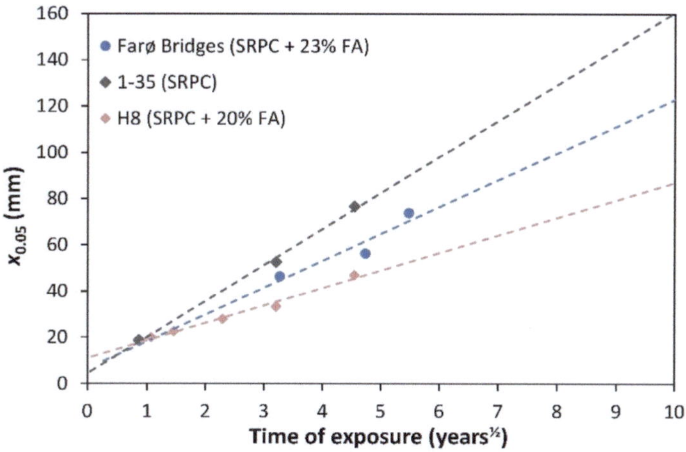


**Fig. 6.9** Application of square root method to Farø bridges and exposure specimens [23]. $x_{0.05}$ refers to the penetration depth corresponding to the chloride concentration of 0.05% mass of concrete, the corresponding $a_{Cr}$ and $b_{Cr}$ are 12 mm/year$^{0.5}$ and 7 mm for Farø bridges. Reproduced from [23] under the terms of the Creative Commons [CC BY 4.0]. © [2024] [Springer]

and ease of use. Machine learning models offer significant advantages in predicting chloride ingress with ability to handle large and complex datasets. However, the effectiveness is highly dependent on the quality and quantity of data, apart from considerable computational resources and expertise. Thus, such models should be used in conjunction with traditional methods and expert judgment to ensure robust and reliable outcomes.

### *6.2.5 Implications for Infrastructure Management*

Long-term exposure practices, as demonstrated through the Danish field exposure sites and studies on old bridges, provide invaluable insights for managing the durability of large-scale structures and infrastructure. These practices offer several key benefits:

- Real-world data: field exposure sites like Rødbyhavn and studies on old bridges provide data on concrete performance under actual environmental conditions, capturing complex interactions that laboratory tests cannot replicate.
- Model validation and refinement: long-term exposure data allows for the validation and refinement of service life models, such as the Femern model, improving their accuracy and reliability for predicting chloride ingress in marine environments.
- Material performance assessment: these studies enable the evaluation of different concrete mixes and their long-term durability, informing better material selection for future projects.
- Chloride threshold determination: field exposure sites with embedded sensors, like those in Rødbyhavn, facilitate the determination of site-specific chloride threshold values, crucial for accurate corrosion initiation predictions.
- Improved design strategies: insights gained from long-term exposure studies inform more effective design strategies for marine structures, potentially extending their service life and reducing maintenance costs.
- Enhanced maintenance planning: by understanding long-term degradation patterns, infrastructure managers can develop more efficient and cost-effective maintenance strategies.
- Advancement of predictive tools: the data collected support the development of advanced predictive tools, including machine learning models, which can further enhance our ability to forecast structural performance.

In conclusion, long-term exposure practices bridge the gap between laboratory research and real-world application, providing a solid foundation for managing the durability of large-scale marine structures throughout their life cycle. This approach not only improves the accuracy of service life predictions but also contributes to more sustainable and resilient infrastructure design and management.

## 6.3 Exposure Site Punta Matamoros at Cayo Santa María (Cuba)

### *6.3.1 Background to the Exposure Site*

The investigations supporting the development and further introduction of the Limestone Calcined Clay Cement (LC3, www.lc3.ch) [6, 44] prompted for the assessment of durability of concrete made with the novel binder composition under real conditions. An industrial trial was carried out for the manufacture of such formulation and concrete blocks were cast [49].

The exposure site is located in the central region of Cuba, specifically in the northern areas of Villa Clara, in Punta Matamoros, Cayo Santa María municipality of Caibarién. Figure 6.10 presents the location where the site was approved and how

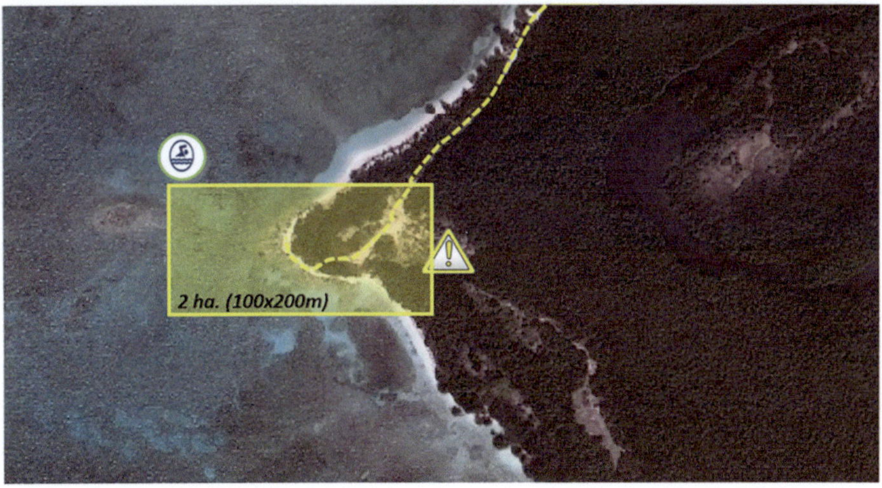


**Fig. 6.10** Location of the exposure site Punta Matamoros (Cuba). Courtesy of Prof. José Martirena-Hernández

to access it. It consists of two exposure platforms with an approximate area of 40 $m^2$, cf. Fig. 1.2. This zone corresponds to the most aggressive exposure class with regards to chlorides, i.e. less than 500 m from the seashore.

In the initial plan, the exposure site included a splash zone and an air born chloride zone. However, the topography of the region made it difficult to have a significant height difference and instead an intertidal area was chosen. Two concrete slabs were built to accommodate the concrete specimens, as shown in Fig. 6.11. Construction works started in 2013 and in the first quarter of 2014 the site was finalised, and the first specimens were laid.

The initial idea of exposure materials was to prepare concrete with Portland cement and with LC3. The LC3 substituted OPC on a one-to-one basis to produce concretes with strength ranging between 20 and 25 MPa with cement content ranging from 300 to 350 kg/$m^3$concrete. Specimens were cast in existing moulds for prefabricated foundations. The original mix designs used are given in Table 6.7.

## *6.3.2 Durability Protocols to Support Management*

The plan was to core the concrete blocks at least once a year and carry out an experimental program to assess the overall durability. The following tests were considered:

- Air permeability according to the Torrent Method [48],
- Surface resistivity [1],
- Porosity, according to NC 345-2011 [17],

**Fig. 6.11** Concrete slabs prepared at the exposure site, Punta Matamoros, Cuba. Courtesy of Prof. José Fernando Martirena-Hernández

**Table 6.7** Mix designs for concrete prefabricated foundations (content in kg/m$^3$)

| Proportioning/Property | Mix design 1 | | Mix design 2 | |
|---|---|---|---|---|
| Design strength (MPa) | 20 MPa | 25 MPa | / | / |
| Ordinary Portland cement | 305 | 325 | 330 | 360 |
| Sand 1 | 592 | 604 | 807 | 780 |
| Sand 2 | 254 | 259 | / | / |
| Gravel | 954 | 973 | 1069 | 1034 |
| Water | 180 | 160 | 155 | 169 |
| Superplasticiser | 3.102 | 3.305 | 3.7 | 4.037 |
| Design water/cement ratio | 0.59 | 0.49 | 0.47 | 0.47 |
| Effective water/cement ratio | 0.50 | 0.40 | 0.40 | 0.40 |

- Rapid Chloride Permeability Test (RCPT) according to ASTM C 1202 [8];
- Chloride profiling [7];
- Carbonation depth according to NC 355: 2004 [16].

The Torrent air permeability method [48] was used to measure the surface air permeability of concrete and have an estimate of damage due to external attack. For the exposure site this method was an appropriate tool to assess surface wearing of concrete due to sea water action. The results in Fig. 6.12 show that the air permeability of OPC concrete increased with time mainly because of the surface deterioration, while LC3 concrete appeared not to be affected by the action of sea water with time [36].

Measuring the surface resistivity of concrete enables assessment of microstructure and transport properties, and the measurement as effectuated through a four-point Wenner probe [1]. Figure 6.13 presents the values of surface resistivity measured in

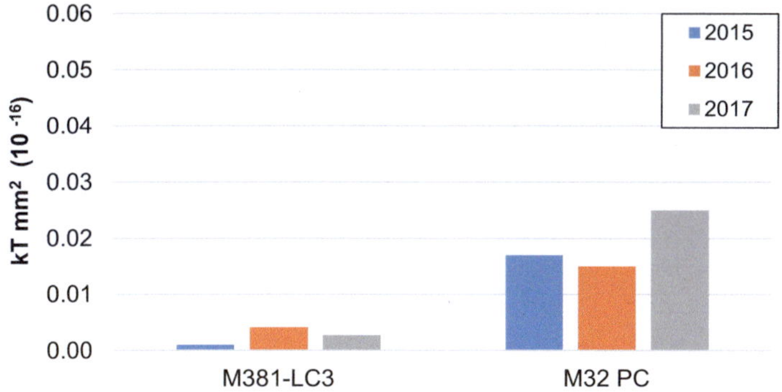


**Fig. 6.12** Values of air permeability on the surface of concrete specimens. This image was originally published in [36] and is licensed under a Creative-Commons license CCBY 4.0

cores taken from the concrete blocks at the exposure site. As expected, the resistivity increased with time, indicating the densification of concrete matrix. This change was much higher for concrete containing LC3, which had a very dense and poorly connected capillary pore network [36].

The porosity test is used to determine the connected porosity of a concrete specimen from measuring the water absorption. For this, a cylindrical specimen of 150 mm diameter and 300 mm height cured at least 28 d or an extracted concrete core from construction element is taken and a thin disc of 25 mm ± 0.5 mm thick is taken from the centre of the specimen (core). The disk is pre-conditioned by oven-drying at 60 °C to constant weight before the water absorption test. The chloride profiling test is conducted according to ASTM C1152/C1152M [7] measuring the acid-soluble chlorides. These tests can be coupled with RCPT according to ASTM C1202 [8] and

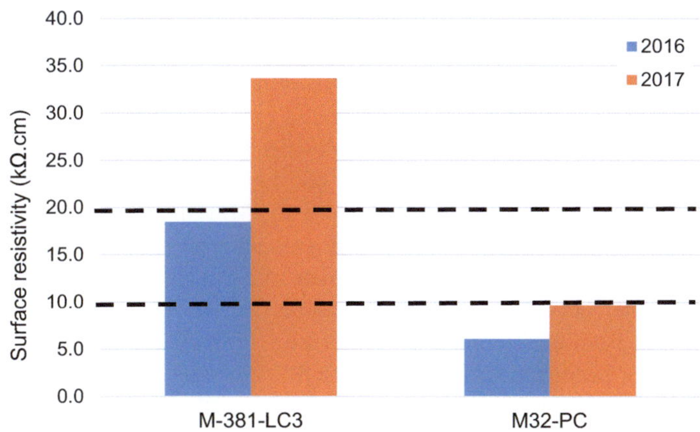


**Fig. 6.13** Measured surface resistivity in concrete samples under natural exposure. This image was originally published in [36] and is licensed under a Creative-Commons license CCBY 4.0

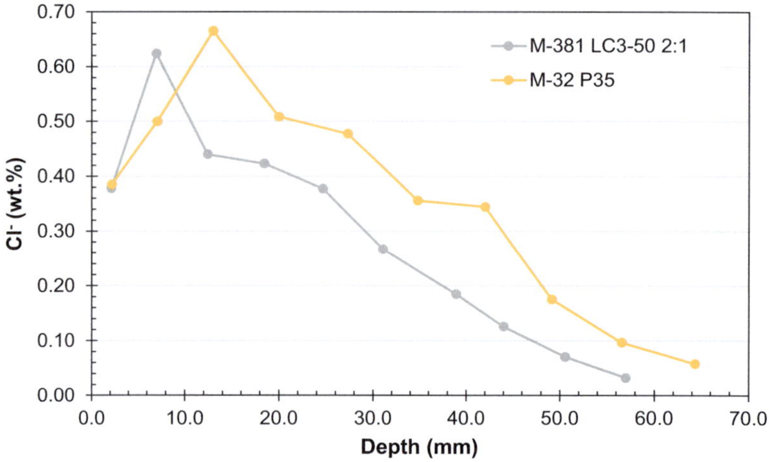


Fig. 6.14 Chloride profiles of concrete cores taken from elements on the site. This image was originally published in [36] and is licensed under a Creative-Commons license CCBY 4.0

give an idea of the resistance to chloride penetration. Figure 6.14 presents chloride profiles measured on concrete specimens with LC3 and OPC, and the concrete with LC3 shows a much lower chloride contents compared to OPC concrete.

### 6.3.3 Validation of LC3 Cement for Exposure Class 1 (Within 500 m from the Seashore)

The investigation was carried out to assess the LC3 cement formulation for concretes subjected to different levels of atmospheric and chemical aggressiveness, following procedures of the Cuban Standard NC-120:2014, 'Hydraulic Concrete—Specifications' [18]. The main aim of the study was to validate the use of LC3 cements with 50% clinker as binder for concrete in highly aggressive conditions [37].

Three series of concretes were prepared using TAC-35 and P35 cements: (H1) 430 $kg/m^3$ of cement and w/c = 0.40; (H2) 405 $kg/m^3$ of cement and w/c = 0.45, and (H4) 345 $kg/m^3$ of cement and w/c = 0.55. Mortars were made in parallel, which were studied in laboratory. Specimens of concretes produced were placed in exposure sites that represent the conditions described in the NC-120:2014 standard [18], which define "very high" as less than 500 m from the coast, "high" as between 500 and 1500 m from the coast and "low" as more than 20 km from the coast.

The concretes produced using LC3 and OPC cements, with the same dose of cement, provide similar compressive strength results at all ages for the same w/c ratio, even though the P35 cement showcased 20% higher resistance at 28 d than the TAC-35. Tables 6.8 and 6.9 present the mix design, the fresh and mechanical properties of the concrete series produced. Figure 6.15 presents the results of mechanical tests.

**Table 6.8** Concrete mix proportioning (kg/m$^3$) of P35/TAC-35 concrete [20]

| Mix type | Exposure class NC 120:2014 | Strength MPa | w/c ratio | c kg/m$^3$ | Aggregate content (kg/m$^3$) | | |
|---|---|---|---|---|---|---|---|
| | | | | | Fine | Medium | Coarse |
| H1 | Very high | 35.0 | 0.40 | 430 | 634 | 352 | 775 |
| H2 | High | 30.0 | 0.45 | 405 | 651 | 362 | 796 |
| H4 | Low | 20.0 | 0.55 | 345 | 690 | 384 | 844 |

**Table 6.9** Compressive strength results of P35/TAC-35 concrete mixes (slump 20 cm)

| Series | Cement kg/m$^3$ | w/c ratio | P35 strength (MPa) | | | TAC-35 strength (MPa) | | |
|---|---|---|---|---|---|---|---|---|
| | | | 3 d | 7 d | 28 d | 3 d | 7 d | 28 d |
| H-35 | 430 | 0.40 | 23.3 | 30.0 | 34.0 | 14.0 | 20.6 | 36.3 |
| H-30 | 405 | 0.45 | 17.3 | 23.0 | 28.0 | 10.3 | 16.3 | 27.0 |
| H-20 | 340 | 0.55 | 12.0 | 14.7 | 19.7 | 7.0 | 12.0 | 18.6 |

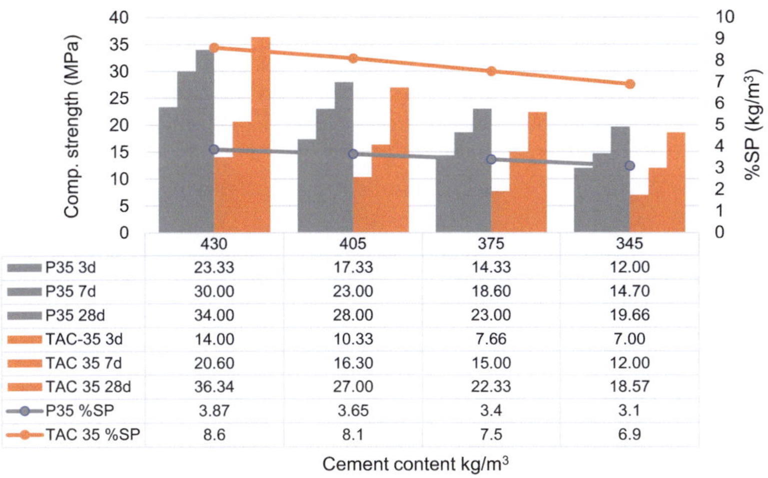


**Fig. 6.15** Results of compressive strength and superplasticiser (SP) demand for all concretes tested [37]. Courtesy of Prof. José Fernando Martirena-Hernández

The studies were carried out in mortars and concretes, at the ages of 6, 18 and 24 months of exposure. The work was oriented to (i) the impact of curing on the carbonation of mortars and concretes exposed in conditions favourable to the occurrence of carbonation, (ii) relation between the material's compressive strength and the changes in the pore structure, evaluated using different methods, (iii) entry and migration of chlorides, both in mortars under pure diffusion conditions, and in concretes with accelerated tests, and (iv) carbonation of mortars and concrete in different environments. The conducted studies included carbonation depth, air permeability, chloride ion migration, electrical resistivity, and effective porosity.

The wet curing of the mortar and concrete series shows a marked influence on the reduction of carbonation at 6 months under favourable conditions in the series produced with TAC-35 cement, while the series produced with P35 has practically no influence. Wet curing also has a marked influence on the densification of pores of concretes produced with TAC-35, especially at 28d. All this shows the importance of curing when working with blended cements.

The relationship between the decrease in compressive strength in concrete with the reduction of the cement content and the increase in w/c ratio was similar in both series (TAC-35 and P35), but the concretes produced with TAC-35 increases the density and decreased connectivity of its pore system, marked by the reduction in water absorption, the increase in surface resistivity and the decrease in air permeability. This may be related to the increased capacity of calcined clay and limestone systems to occupy more efficiently the pore spaces for the precipitation of reaction products. Mortars produced with TAC-35 immersed for 6, 18 and 24 months in a 3% NaCl solution showed a significant reduction in the total chloride content and penetration depth, compared to mortars produced with P35 treated under the same terms, cf. Fig. 6.16. For the same depth, e.g. 20 mm, the concentration of chlorides in P35 series is 7 times higher than TAC-35 series. The results of the apparent diffusion coefficient of P35 series are 9–11 times higher than TAC-35 series. This trend is consistent with the dense and poorly connected network of pores in the matrix of TAC-35, which slows down the entry and migration of ions.

The chloride permeability of all series of concrete produced with TAC-35 cement, measured from the Rapid Chloride permeability test (ASTM 1202) is 12–15 times higher than that of the series produced with TAC-35. These results are very consistent with the relationships obtained between the apparent diffusion coefficient, measured

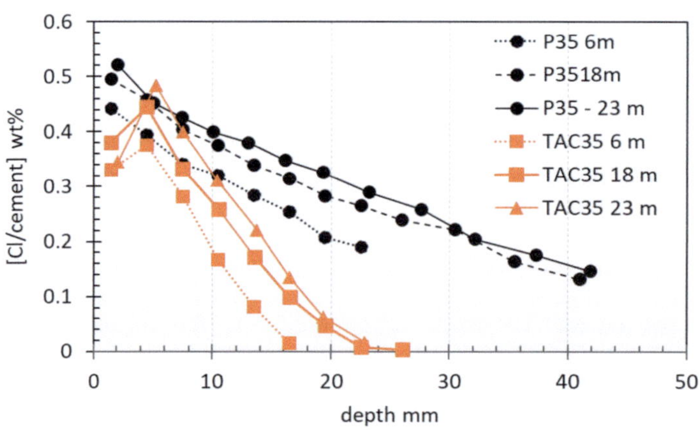


**Fig. 6.16** Results of chloride profiling in mortars submitted to pond exposure [37]. *Source* Courtesy of Prof. José Fernando Martirena-Hernández

in mortars produced with TAC-35 and P35. The measurement of the carbonation profiles in mortars and concrete at the ages of 6, 18 and 24 months confirms the trend of higher carbonation in the systems produced with TAC-35, when compared with the results of the series with P35. In conditions of high chloride aggressiveness, for example, in coastal areas, the carbonation depths of the series with TAC-35 are in the same range of the series produced with P35. This occurs due to the high relative humidity (average higher than 70%), where the pore system of the concrete is saturated, therefore, it is very difficult for the $CO_2$ to enter and dissolve in the matrix. In the rest of the series, with an exposure far from the coast, the relative humidity decreases to levels that oscillate between 50 and 60%, and the gas–liquid interface is favoured in which the environmental $CO_2$ can dissolve and penetrate the cementitious matrix. However, in these systems the presence of chlorides is markedly lower, so the danger of corrosion by chlorides is lower.

This study confirms the advantages of using LC3 type cements in the manufacture of structural concretes in highly aggressive areas, especially in coastal areas with high concentrations of chlorides.

### *6.3.4 Validation of Concrete with Recycled Aggregates for Use in Marine Environments*

The investigation focused on assessing the durability of a concrete made with treated recycled aggregates. A new treatment method of recycled aggregate, called encapsulation or coating, has been used to enhance the properties. The durability testing program compared the behaviour of different recycled aggregate concretes with natural aggregate concrete (NAC) under three environmental conditions. The pore structure and migration properties of concretes have been assessed during up to 30 months of exposure, by using several durability tests. Test results showed a viable possibility for concretes entirely produced with recycled concrete aggregates, when the encapsulation treatment is applied. Concretes which have been produced in May of 2016, labelled as RAC-E 2016 and NAC, were cured for 28 d under water and exposed to the three sites in October 2016. Durability test series were proceeded after 5, 17 and 30 months of exposures.

To get an ampler investigation, a second series of concretes, labelled as RAC-E 2017 and RAC, were produced in May/June of 2017. Durability tests of these specimens were conducted after 9 and 22 months of exposure. Figure 6.17 shows the protocol used to assess the durability in concrete in time sequences, including concrete production and the time of exposure.

Cores with diameters of 50 and 100 mm, as shown in Fig. 6.18, were retrieved at the designated times and were sealed at the site to prevent changes in moisture until the durability tests, including electrical resistivity, carbonation under natural and accelerated conditions, water absorption, rapid chloride migration test and chloride profiling. With the designed experimental program, it has been proved, that the

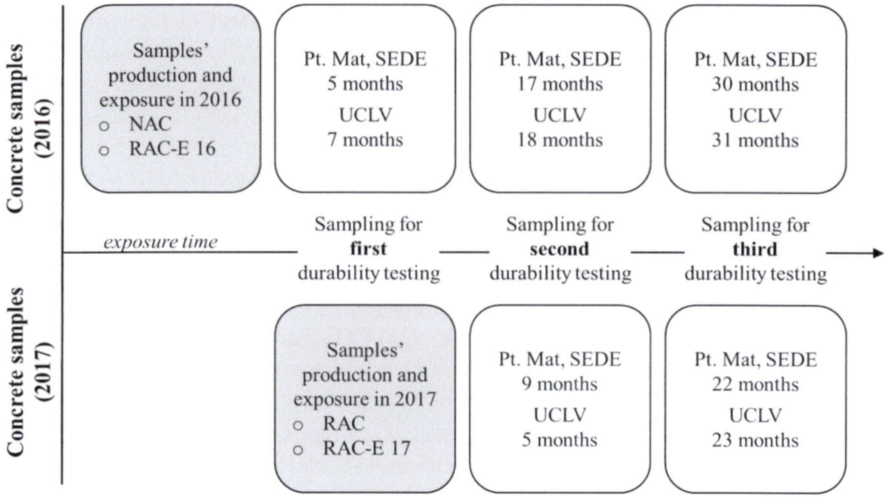


**Fig. 6.17** Graphical representation of the experimental program

encapsulation treatment of the coarse fraction from 5 to 10 mm of recycled concrete aggregates has a positive effect on the physical properties of the aggregates such as water absorption, and also on concrete properties of 100% recycled aggregate concrete [34].

**Fig. 6.18** Concrete specimens at the exposure site. Courtesy of Prof. José Fernando Martirena-Hernández

**Table 6.10** Concrete mix design for elements of the wharf structure (C for cement, FA for fine aggregates and CA for coarse aggregates)

| Concrete element | Strength class | Cement (kg) | w/c ratio | C: FA: GA |
|---|---|---|---|---|
| Square block | C23 | 340–380 | 0.47–0.50 | 1:1.731:2.91 |
| Wharf rail | C28 | 350–384 | 0.41–0.45 | 1:2.02:3.295 |
| Support beam | C48 | 400–480 | 0.38–0.40 | 1:1.70:2.54 |
| Prestressed T-beam | C48 | 420–475 | 0.38–0.40 | 1:1.46:2.30 |
| Pipe channel beam | C32 | 340–355 | 0.46–0.47 | 1:2.10:2.96 |
| Support beam | C28 | 340–375 | 0.46–0.50 | 1:1.90:2.74 |
| Elevated part for support beam | C33 | 335–355 | 0.46–0.47 | 1:2.24:3.10 |

**Table 6.11** Chemical composition of seawater for R port (mg/l)

| Ion species | $K^+$ | $Na^+$ | $Ca^{2+}$ | $Mg^{2+}$ | $Cl^-$ | $HCO_3^-$ | $CO_3^{2-}$ | $SO_4^{2-}$ |
|---|---|---|---|---|---|---|---|---|
| Concentration | 437 | 8,910 | 326 | 1,200 | 16,378 | 165 | 33 | 2,495 |

## 6.4 Concrete Structure of R Coal Wharf

### *6.4.1 Background*

The R Port contains a 100,000-tonnes coal wharf, built in 1985 as a gravity pier-type wharf on the shore of the Yellow Sea, China. After more than ten years of operation, the upper structure of the wharf showed durability deterioration related to the corrosive environments. Commissioned by the owner, an extensive inspection program was installed in April 2002 to evaluate the durability and safety of this concrete structure. Since the inspection was performed on the original concrete elements (cores), the measured properties can be regarded as a special sort of exposure data. It is from this angle that this wharf structure inspection is included. The inspection program covered the carbonation depth, the chloride ingress, the thickness of concrete cover, the state of corrosion, and the strength of concrete. Table 6.10 provides some preliminary information for the concrete mixture proportioning.

To characterise the environmental actions, the composition of the seawater, collected from the flushing procedures, is analysed and the results are provided in Table 6.11.

### *6.4.2 Inspection and Durability Testing*

The scope of the inspection focused on structures above elevation 8.0 m, and reported on chloride ingress, carbonation, and corrosion of reinforcing steel rebars. The revealed deterioration included exposed steel by concrete spalling, corrosion of

**Table 6.12** Concrete carbonation test results

| Concrete element | Location and position | Average carbonation depth (mm) |
|---|---|---|
| Square block | Wharf No.4 square block | 2.0 |
| | Wharf No.6 square block | 1.3 |
| Support beam | Wharf 3 north | 1.0 |
| | Wharf 6 north | 3.2 |
| | Wharf 6 south | 1.4 |
| | Wharf 7 north | 1.4 |
| | Wharf 7 south | 1.2 |
| Wharf concrete beam | Wharf 4 west 1# | 2.4 |
| | Wharf 4 west 2# | 1.8 |
| | Wharf 6 west 1# | 6.2 |
| | Wharf 6 west 2# | 9.0 |
| 32 m prestressed concrete beam | Between wharf 5 and wharf 6 | 2.0 |
| | Between wharf 6 and wharf 7 | 2.0 |

reinforcement, exposure of prestressed tendons, and notable cracking. The main inspection and the related results are reviewed below.

#### 6.4.2.1 Carbonation Depth

Concrete carbonation depth was detected using a phenolphthalein solution [35]. The scope of this test included the support beam, the ordinary concrete beam of the pier, the 32 m prestressed concrete beam, and the square block. The carbonation depth results are given in Table 6.12.

The carbonation resistance of concrete is directly related to the compactness of concrete surface. From the external inspection, it was found that the surface condition of concrete was relatively good, and the depth of carbonation of the concrete was small, with the maximum depth of carbonation reaching 11 mm and the most depths of carbonation ranging from 1 to 9 mm. Considering the significant thickness of concrete covers, e.g. more than 50 mm, the carbonation was not responsible for the corrosion of reinforcement in concrete structures.

#### 6.4.2.2 Chloride Penetration into Concrete

The inspection scope for chloride penetration covers square blocks, support beams, longitudinal beams and support beam. Concrete cores were taken from the beams and sliced every 10 mm along the depth direction, the mortar was made into powder after removing the coarse aggregate and the $Cl^-$ content was detected in it by photometric method [35]. The limit values of chloride content are prescribed as 0.06% and 0.10% of binder mass ratio for prestressed and reinforced concrete elements by the Chinese standards [14, 15]. Figure 6.19 shows the profiles of chloride penetration into different elements.

From the figure, it can be seen that all the four elements suffered from chloride penetration and the surface chloride content is much higher than the internal one. For the square block situated in SPL zone, the penetration depth was more than 50 mm, equal to the concrete cover thickness for rebars. The chloride penetration depth for support beams in ATM zone reached also 50 mm and locally to 80 mm, due to local damage of concrete surface by water flows. The 32 m prestressed beam, exposed

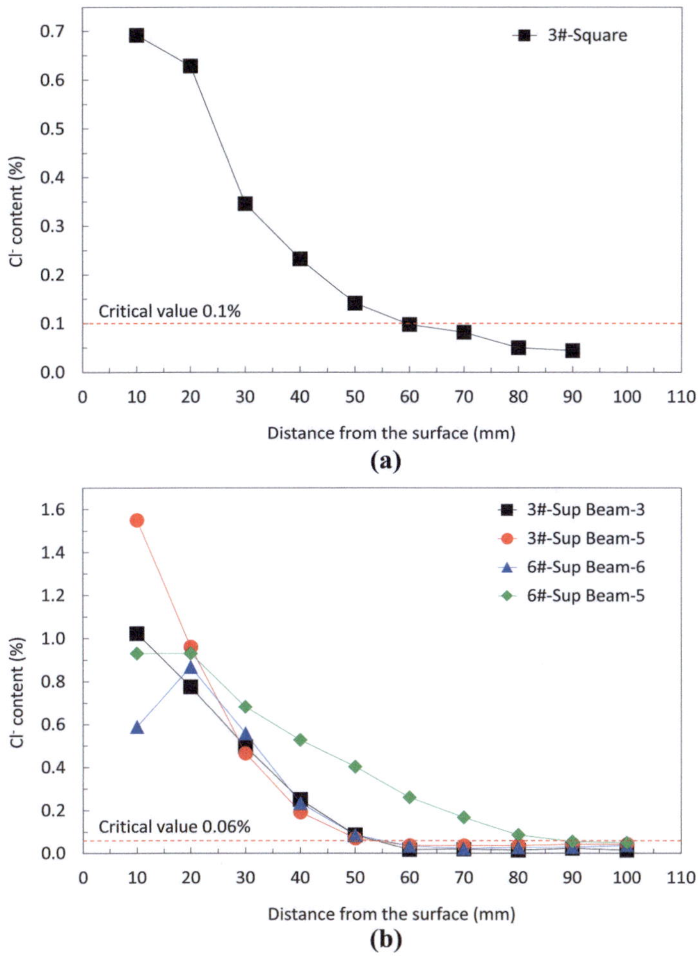


**Fig. 6.19** Chloride profiles for **a** square blocks, **b** support beam, **c** 32 m prestressed beam and **d** 11 m longitudinal beam

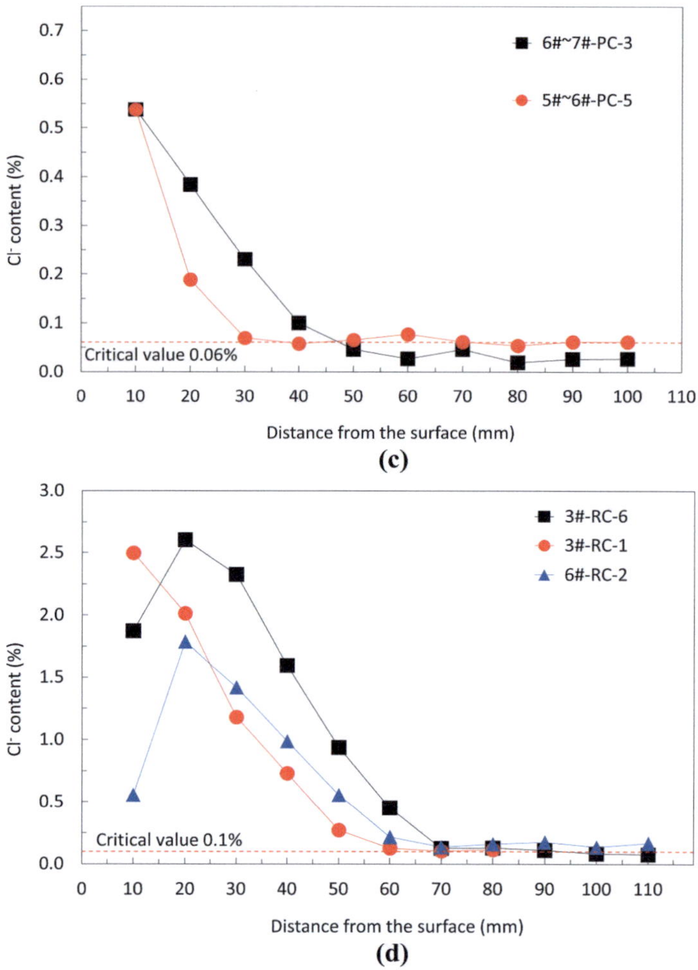


**Fig. 6.19** (continued)

in ATM zone, had chloride penetration depth, with 0.1% as limit value, attained 30–40 mm, leading to local corrosion when the nominal concrete cover thickness 50 mm was not respected. The longitudinal beams, exposed in ATM zone, had the most advanced penetration depths, in the range of 80–100 mm, due to the relative low grade of structural concrete (C28). The particular feature of the longitudinal beam is the stabilised chloride content, deeper than 70 mm, remained at 0.10–0.14%, still higher than the limit of 0.1%. It is possible that the raw materials of concrete had been contaminated by external chlorides during the construction. Note that these Chinese standards adopt stricter limits on chloride threshold content, e.g. compared to other standards such as ACI-318 [2].

### 6.4.2.3 Corrosion of Reinforcement

The inspection of reinforcement corrosion mainly focuses on the north bearing girders and the prestressed reinforced concrete girders. Both non-destructive testing and destructive testing were used to detect the corrosion of steel bars. The non-destructive testing refers to the electrochemical methods including the measurements of half-cell corrosion potential [19], corrosion current density [19] and resistivity of concrete [19], and the destructive test consists of removing the concrete in corrosion-suspected areas and checking the corrosion state of steel bars directly.

Extensive tests were performed on one of the support beams (North No.3) for the corrosion potential, corrosion current and resistivity. The half-cell potential results are showed in Fig. 6.20 and Table 6.13 summarises the corrosion current and resistivity.

The corrosion potential results in Fig. 6.20 are lower than −200 mV (absolute value) in most areas. The possibility of rebar corrosion in the concrete is higher, the concrete in the place of high corrosion potential of rebar was then chiselled to expose the rebar therein, and the inspection found that the rebar was indeed corroded. A comparison between the potential in Fig. 6.20 and the inspection showed that the lower potential areas correlated well with the concrete cracking due to corrosion products. Table 6.13 gives the corrosion current density and concrete resistivity, showing a moderate corrosion rate. Figure 6.21 shows the heavily corroded reinforcement and the concrete cracking in the north support beam at the joint height level. This heavy corrosion was attributed to the facts that the support beam had been extended horizontally by C28 and vertically by C33 and cracking occurred between the two concretes. This cracking due to construction measures promoted the subsequent chloride penetration and the corrosion of reinforcement.

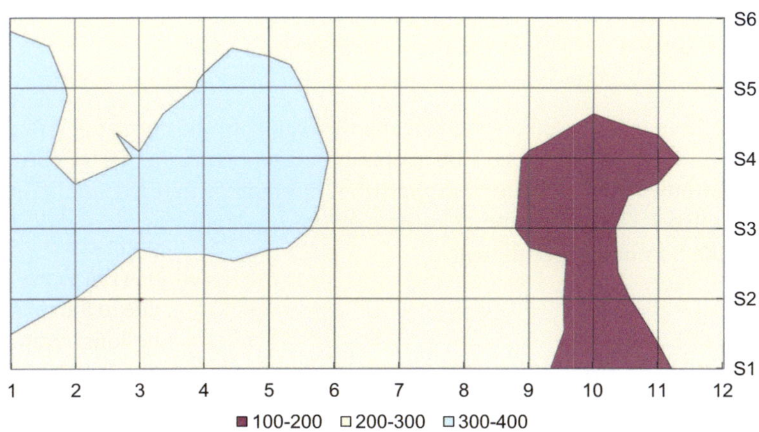


**Fig. 6.20** Corrosion potential measured on the surface of No.3 north support beam. The half-cell potential values should take the negative of illustrated values

**Table 6.13** Measurements of corrosion current density and resistivity for support beam

| Column | Measurements | Line S2 | Line S4 | Line S6 |
|---|---|---|---|---|
| 1 | Corrosion current density (μA/cm$^2$) | 0.500 | 0.21 | 0.047 |
| | Resistivity (kΩ.cm) | 17.17 | 26.24 | 15.82 |
| 3 | Corrosion current density (μA/cm$^2$) | 0.124 | 0.061 | 0.21 |
| | Resistivity (kΩ.cm) | 20.39 | 46.01 | 11.41 |
| 5 | Corrosion current density (μA/cm$^2$) | 0.139 | 0.088 | 0.033 |
| | Resistivity (kΩ.cm) | 25.93 | 33.42 | 45.56 |
| 7 | Corrosion current density (μA/cm$^2$) | 0.087 | 0.127 | 0.064 |
| | Resistivity (kΩ.cm) | 36.12 | 25.87 | 27.24 |
| 9 | Corrosion current density (μA/cm$^2$) | 0.096 | 0.65 | 0.078 |
| | Resistivity (kΩ.cm) | 27.31 | 23.02 | 31.28 |
| 11 | Corrosion current density (μA/cm$^2$) | 0.048 | 0.191 | 0.076 |
| | Resistivity (kΩ.cm) | 21.51 | 28.90 | 30.53 |

**Fig. 6.21** Horizontal cracking caused by corrosion (left) and confirmed corroded reinforcement after concrete removal. *Source* Courtesy of Dr. Tingyu Hao

For the 32 m prestressed concrete beam, the results of the corrosion potential are given in Fig. 6.22. Based on the values of half-cell potential, the risk of corrosion of rebars was rather low, knowing that the strength grade of the structural concrete is C48. Also, from Fig. 6.19c, the chloride penetration depth was around 40 mm, far less than the nominal cover thickness.

### 6.4.3 Support for Management

The R port case study provides an example of how long-term durability data, in form of in-situ inspection and testing results, support the maintenance and management of concrete infrastructures in marine environment. After 17 years of operation, the

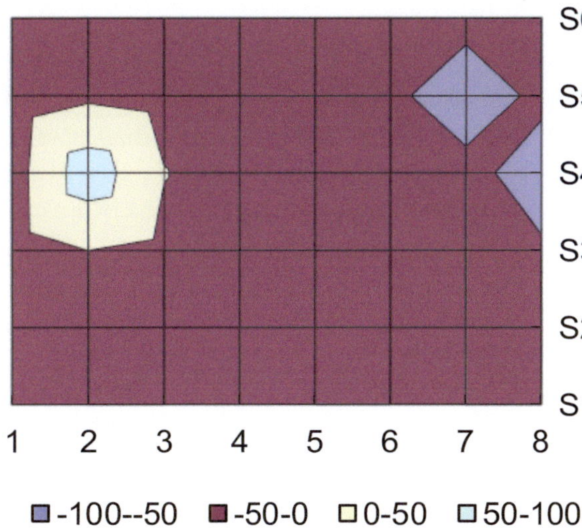


**Fig. 6.22** Corrosion potential of 32 m prestressed concrete girders between No. 5 and No. 6 wharfs

100,000-tonnes coal wharf structures suffered deteriorations due to chloride penetration, including corrosion of rebars, concrete cracking and spalling, which poses a threat to both the durability and the safety of the structures. The inspection carried out in 2002 helped to reach the following conclusions.

1. The chloride ingress is the primary cause of damage for most of the structural elements on the upper part of the wharf (>+8.0 m). In the daily operations, measures should be put in place to protect the structural elements from the direct splashing of seawater during the flushing practice.
2. The chloride content in the concrete cover of the square blocks, about 50 mm inward from the concrete surface, exceeds the limit of 0.1%, which, considering the exposure to SPL zone, caused serious corrosion of the reinforcement. This is the main cause of the concrete spalling at the corners and subsequent exposure of the reinforcement.
3. Due to the measures taken in the construction for support beams, cracking appeared already between the different concretes. Such cracking had an accelerating effect on the corrosion of the steel reinforcement, and these cracks should be repaired separately.
4. Few of the 32 m prestressed girders between the spans of the wharf showed signs of corrosion, and the depth of chloride ingress is in the range of 30–40 mm. Nevertheless, corrosion risk was still a tangible risk, and anti-corrosion measures, such as electrical chloride extraction, would be necessary.
5. Reinforced concrete longitudinal beams on double pier wharf, the internal chloride content was in the range of 0.10–0.14%, the external chloride penetration depth was around 70 mm, considering its low strength grade (C28), the corrosion risk of steel reinforcement was realistic and comprehensive maintenance of reinforcement rebars against corrosion became necessary.

**Fig. 6.23** Two exposure sites for studying marine durability of concrete in Spain: Barcelona in the Mediterranean Sea and Huelva in the Atlantic Ocean. *Source* Courtesy of Dr. Carmen Andrade

## 6.5 Exposure Sites in Spain

### *6.5.1 Introduction*

In Spain two main exposure sites for evaluating concrete durability in sea water were designed. The first one in Mediterranean Sea was established in 1982 in Barcelona to study the corrosion rates of reinforced concrete in real conditions; and the second was built in 1988 in Huelva, in the Atlantic Ocean, to study the attack of seawater to reinforced concrete with different cement types and its rebars' corrosion rates, as shown in Fig. 6.23.

Both sites had the objective to gain understanding on the reinforcement corrosion process by studying the values of the instantaneous corrosion current, corrosion potential and resistivities achieved in the different stages of the propagation phase. Also, the values between the two sites were compared to find out the influence of the type of cement, the concrete mix proportions and the cover depth on the corrosion parameters. Additionally, in Barcelona station the behaviour of the normal reinforcement was compared with galvanised bars and the protective effect of calcium nitrite added in the mix as corrosion inhibitor was investigated. The following will briefly describe both exposure sites, the variables studied, and the lessons learned for the long-term corrosion management for concrete structures in marine environments.

### *6.5.2 Exposure Site in the Mediterranean Sea (Barcelona)*

The facility was built in 1982 at the Institute of Sea Sciences of the High Research Council of Spain (CSIC) in Barcelona. This institute has a large aquarium with a system to pump natural seawater from the Mediterranean Sea continuously into the tanks where many marine species live. Two tanks were built specifically for the corrosion study (Fig. 6.24). The water circulates continuously and was filtered for avoiding any contamination from organics and biofouling. The tanks were placed in the basement of the building and maintained at rather constant temperature. The mean salinity of the water was around 38 g/l, the temperature ranged between 14 °C (winter) and 24 °C (summer), and the oxygen content was between 5 and 5.71 ppm.

The concrete specimens had a size of 15 × 20 × 10 $cm^3$, as shown in Fig. 6.25. The concrete specimens were moisture-cured in a chamber at nearly 100% relative humidity for 60 d and then transported to Barcelona and placed in the tanks. Each sample had ten reinforcing steels of 8 mm in diameter, embedded at various depths to have different cover depths: 0.5, 1.5, 2.5, 4.0 and 7.5 cm. The lateral surfaces of the specimens were covered with epoxy resin and a final wax in order to allow the penetration of chlorides only through the front and back surface.

The concrete mix proportions are reported in Table 6.14. The variables tested were: i) ordinary steel for reinforcing of 8 mm in diameter with nominal tensile strength of 400 MPa and galvanised steel; ii) addition of sodium nitrite as corrosion

**Fig. 6.24** Natural sea water tank (left) with specimens submerged (right) in the Institute of Sea Sciences of the High Research Council of Spain (CSIC) in Barcelona, Spain. *Source* Andrade [4], Fig. 2. Reproduced with permission of Elsevier

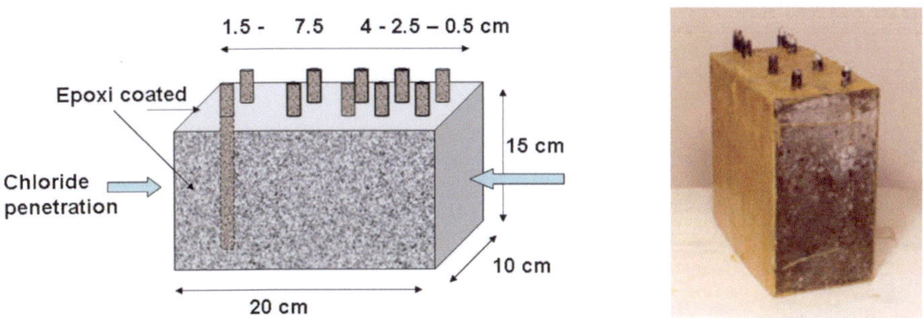


**Fig. 6.25** Dimensions of the concrete specimens and cover of the reinforcement (left) and cast concrete specimen (right). *Source* Andrade [4], Fig. 1. Reproduced with permission of Elsevier

**Table 6.14** Mix proportions of the concrete and characteristics

| Cement ($kg/m^3$) | Coarse aggregate ($kg/m^3$) | Sand ($kg/m^3$) | w/c ratio | Slump (cm) | Porosity (%) |
|---|---|---|---|---|---|
| 400 | 1045 | 855 | 0.38 | 1 | 6.85 |
| | 990 | 810 | 0.60 | 14 | 10.01 |

inhibitor; iii) type of cement (ordinary Portland cement, pozzolanic cement, fly ash cement and slag cement), iv) proportions of the concrete mix; and v) cover depth from 1.5 to 7.5 cm.

#### 6.5.2.1 Measuring Techniques

The corrosion potential ($E_{corr}$) and the linear polarisation resistance (LPR) were obtained using a portable potentiostat. The $E_{corr}$ was recorded using a calomel electrode in the first years and then a silver/silver chloride one with close reference potential value to the calomel one. The Linear Polarisation method (LRP) [5] measures the ratio between a small electrical voltage ($\Delta E$) imposed on the metal and the corresponding current response ($\Delta I$), i.e. the polarisation resistance (Rp). A potentiostatic method was used for the small specimens and the waiting time for the response was around 30 s. The ohmic drop due to the concrete was measured simultaneously at the time of switching-off the voltage, which is subtracted from the value of $R_p$. The corrosion rate was then calculated using

$$I_{corr} = \frac{B}{(R_p - R_c) \cdot A} \tag{6.4}$$

where $R_c$ is ohmic or concrete resistance, A is the area exposed to attack, taking 23 $cm^2$, and B is a constant of 26 mV. The measurements were made during the first two months daily, but once the specimens were moved to the sea water tank they were measured only once per year. After different times of exposure, the specimens were broken and the bars were removed, cleaned and weighed to compare the electrochemical and gravimetric losses [3].

#### 6.5.2.2 Results

In Fig. 6.26 is shown a typical behaviour of the specimen made with 400 $kg/m^3$ OPC cement and w/c = 0.6: the evolution of the corrosion current ($I_{corr}$) for 23 years and the cumulative corrosion ($P_{corr}$) in function of the cover depth.

Figure 6.27 shows the comparison of the different variables studied on the case of OPC cement: amount of cement, type of rebar and presence on nitrites as inhibitor [42]. The use of galvanised steel or a corrosion inhibitor significantly extended the

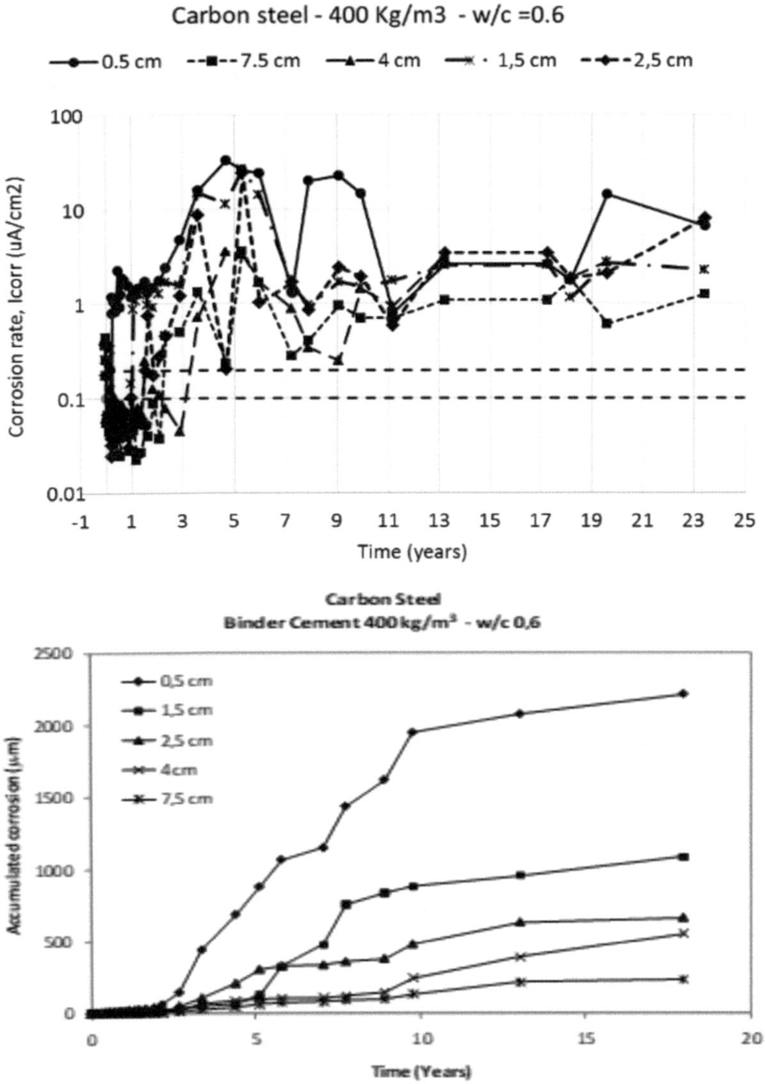


**Fig. 6.26** Influence of concrete cover on potential corrosion (up) and the cumulative corrosion (down). *Source* Adapted from [4], Fig. 8, Fig. 19. Reproduced with permission of Elsevier

depassivation time in all cover depths while the use of a higher w/c, 0.60 compared to 0.52, decreased significantly the depassivation time.

The results indicated that the levels of corrosion current are irrespective of the binder or steel type when corrosion is negligible (i.e., below 0.1 $\mu A/cm^2$). However, the depassivation increased the values by nearly one order of magnitude. Regarding

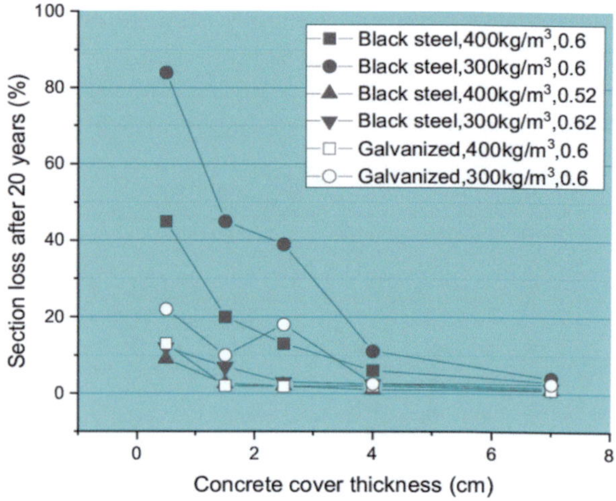


**Fig. 6.27** Comparison of the loss of section of different concrete mix proportions of OPC cement and the impact of using a corrosion inhibitor or galvanised steel [42]

the accumulated corrosion, it should be noted that all variables influencing the corrosion current also affect the total accumulated corrosion.

### *6.5.3 Exposure Site in the Atlantic Ocean (Huelva)*

The exposure station is located in Huelva in the south of Spain, near the frontier of Portugal, at the Atlantic Ocean, as shown in Fig. 6.23. The private beach belongs to the Huelva Harbor Authorities. The site station was created in 1988 to study the resistance to seawater attack of different cement types and further extended to study the reinforcement corrosion. The composition of seawater is $Mg^{2+}$ (2,400 mg/L), $Cl^-$ (19,900–20,200 mg/L), $SO_4{}^{2-}$ (2,900 mg/L) and $Ca^{2+}$ (500 mg/L), the pH value is 6.4–8.1, and the conductivity is 20.2 S/cm. The composition of the steel bars was the same as the previous section and the diameter of the steel bars was 16 mm.

In addition to mortar specimens and cylinders for concrete strength, large concrete blocks of 50 × 50 × 200 $cm^3$ were fabricated and partially buried in the beach in 1989 at 20 and 25 m from the lower tide, as shown in Fig. 6.28. The blocks were reinforced by four main steel bars in the long dimension with a cover depth of 3 cm, which were linked with stirrups in the shorter dimension. A bent reinforcement or two were welded to the main reinforcement as a "hook" for moving and placing of the blocks, (Fig. 6.27), and these hooks had an essential role in the corrosion performance as shown later.

Several types of cement with and without mineral additions were used in this exposure site. The concretes to be monitored for corrosion had a cement content of 200 $kg/m^3$ (normal dosage for blocks in 1988) and 350 $kg/m^3$ with w/c of 0.5 and a slump of 1 cm. As opposed to the specimens for the tank, the blocks were water-cured in the fabrication site and then transported to the beach at age of 28 d. Over years of

**Fig. 6.28** Beach divided by a pier (left), block with the growing of biofouling and hook welded into the main reinforcement (centre), and placement of the blocks (right). *Source* Andrade [4], Fig. 5. Reproduced with permission of Elsevier

seawater exposure, diverse forms of fouling accumulated on all concrete surfaces, yet this layer did not seem to hinder chloride penetration.

#### 6.5.3.1 Measuring Techniques

For the concrete blocks, the LPR was measured using the portable corrosion rate meter GECOR 8 and GECOR10 shown in Fig. 6.29 [21], and the reference electrode was of $Cu/CuSO_4$. The measurements were only effectuated in 2009 (after 21 years) and in 2014 (after 26 years). In addition to the electrochemical measurements, the cover depth was removed to verify the corrosion extent of steel bars due to the chlorides' penetration.

**Fig. 6.29** Measurement of LRP in the large concrete blocks by using the GECOR 10 which confines the current through a guard ring. *Source* Adapted from [4], Fig. 7. Reproduced with permission of Elsevier

#### 6.5.3.2 Results

Corrosion measurements were made in 2000, 2009 and 2014 only in several blocks because the values were very similar as shown in Fig. 6.30. All the values were above 1 $\mu A/cm^2$ and the potential values very negative, below −600 mV ($Cu/CuSO_4$).

These values indicated high corrosion extent and therefore the concrete cover was removed in some places, during inspections in 1994, 2009 and 2014, to observe the rebar condition directly. Surprisingly, the reinforcing bars were completely clean, even in the very superficial zone at the interface between the air and the concrete. The explanation proposed was that the external hook acted as sacrificial anode for the

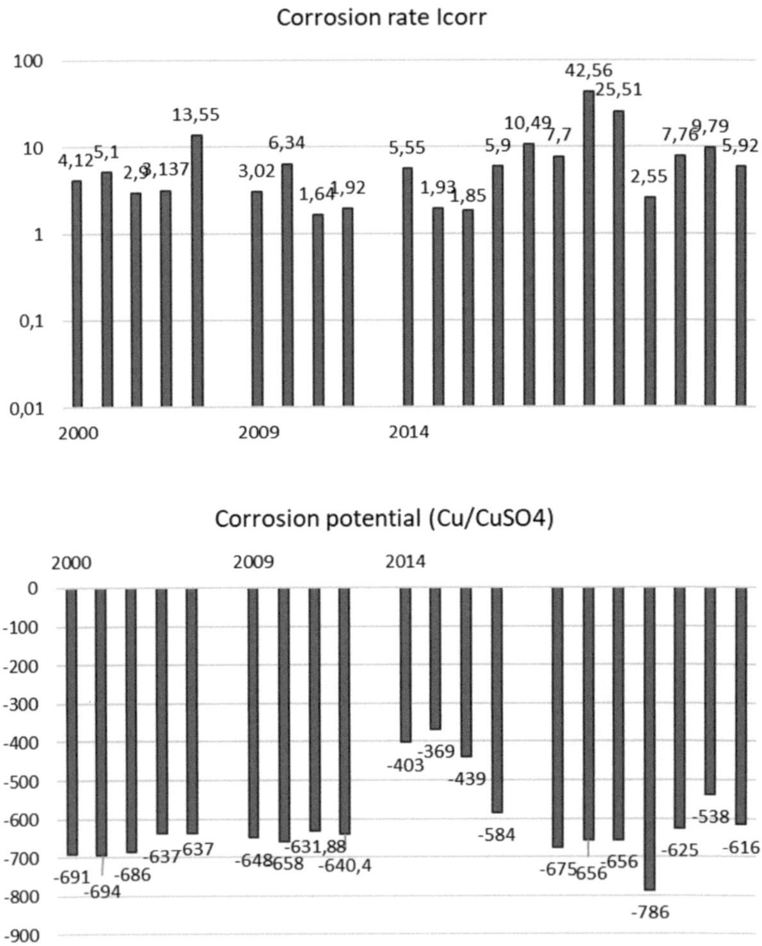


**Fig. 6.30** Electrochemical measurement for the large concrete blocks at the beach: corrosion rates (up) and corrosion potential (down). *Source* Andrade [4], Fig. 10. Reproduced with permission of Elsevier

internal reinforcement, thereby protecting all bars against corrosion [4, 9] despite the high level of chloride penetration. However, it was particularly remarkable to observe that although the bars appeared clean immediately after removing the concrete cover the bars, small pits began to develop over their entire surface after only one hour of air exposure. This delayed onset of corrosion was attributed to the loss of the protection that occurred when the electrolyte, i.e. the concrete cover, was removed This observation confirmed the high degree of chloride contamination and demonstrated that, once the cathodic protection ceased due to the removal of concrete cover, pitting corrosion initiated almost immediately.

### 6.5.4 Exposure

The present work compiles corrosion rate data from structures and specimens subjected to real marine environmental conditions, specifically in the ATM and TID zones. There are relatively few data of real structures in the literature; therefore, the results are highly valuable. The measurements contributed to establishing the ranking incorporated into the RILEM Recommendation [3] and confirmed that the corrosion current depended on the chloride concentration at the steel bar level. Consequently, the polarisation resistance technique proves to be very suitable for determining the instantaneous corrosion current, detecting the passivation state of the reinforcement, and assessing the intensity of the corrosion. The passive state of steel bars is lost when current values permanently exceed above 0.1–0.2 $\mu A/cm^2$. In elements located in TID zone, very high corrosion rates, above 10 $\mu A/cm^2$, are possible, whereas in ATM zone, corrosion activity is not generally lower, with typical maximum values around 1 $\mu A/cm^2$. The accumulated corrosion or corrosion penetration $P_{\mathrm{corr}}$, resulting from the time integration of the corrosion rate $I_{corr}$, provides a clearer distinction of the time to corrosion initiation time in relation to the cover depth or concrete characteristics.

## 6.6 Chapter Summary

This chapter reviewed selected case studies of concrete structures and concrete samples exposed to marine conditions in different areas of the world, including two examples in China, three bridge structures in Denmark, concrete samples exposed to marine conditions in a testing site in Cuba, and a case study of concrete samples in Spain, subjected to marine exposure in the Mediterranean Sea and the Atlantic Ocean. These cases provide rich experiences in exploiting the long-term exposure practice to reveal scientific mechanisms, establish service life models, and make management and maintenance decision as well.

For HZM bridge in China, a systematic framework was established for the management of the long-term durability during the service life, through feeding the exposure data into durability models and updating the maintenance schemes for concrete structures. For the three old bridges in Denmark, the Femern model and square root method were calibrated from the exposure sites and applied to bridge cases, eventually to help the service life prediction the maintenance planning of the bridges in use. Novel binders, such as LC3, were also put into exposure practice in Cuba to investigate their real behaviours compared to the conventional binders. From the exposure practice in Cuba, the results helped to validate the LC3 cement used in typical marine environment and the use of recycled aggregates as well. The coal wharf in R port provides a typical case for reinforced concrete and prestressed concrete structures exposed directly in marine environment, in which the deterioration was correlated directly to the strength grade of concretes. The electrochemical measurements helped to set up the rational maintenance schemes. The exposure results from the Spanish sites are informative for the electrochemical behaviours of different steel bars embedded in concrete and exposed to different environmental actions. The obtained results helped to establish the ranking of environmental actions and the passivation limit for corrosion current, which is rather fundamental for engineering use.

In conclusion, durability data, especially from the exposure practice, play an indispensable role in the management of concrete infrastructure exposed to marine conditions. By providing critical insights into the behaviour of concrete and exploiting the data using reliable durability models, these data can better support structure design, maintenance decision, and prediction of infrastructure lifespan. As a result, the careful collection and application of durability data can significantly enhance the safety, sustainability, and cost-efficiency of marine infrastructure projects, ensuring their functionality and security over the long term.

## References

1. AASHTO: Standard method of test for surface resistivity indication of concrete's ability to resist chloride ion penetration. American Association of State Highway and Transportation Officials (AASHTO). Washington, D.C., USA (2019)
2. ACI: Building code requirements for structural concrete and commentary (ACI 318-19). American Concrete Institute, Farmington Hills, USA (2019)
3. Andrade, C., Alonso, C., Gulikers, J., Polder, R., Cigna, R., Vennesland, Ø., Salta, M., Raharinaivo, A., Elsener, B.: RILEM TC 154-EMC: electrochemical techniques for measuring metallic corrosion. Recommendations test methods for on-site corrosion rate measurement of steel reinforcement in concrete by means of the polarization resistance method. Mater. Struct. **37**, 623–643 (2004)
4. Andrade, C.: Steel corrosion rates in concrete in contact to sea water. Cem. Concr. Res. **165**, 107085 (2023)
5. Andrade, C., Alonso, C.: On-site measurements of corrosion rate of reinforcements. Constr. Build. Mater. **15**, 141–145 (2001)
6. Antoni, M., Rossen, J., Martirena, F., Scrivener, K.: Cement substitution by a combination of metakaolin and limestone. Cem. Concr. Res. **42**, 1579–1589 (2012)
7. ASTM: Standard test method for acid-soluble chloride in mortar and concrete (ASTM C1152/C1152M). ASTM International, US (2020)

8. ASTM: Standard test method for electrical indication of concrete's ability to resist chloride ion penetration (ASTM C1202-22). ASTM International, US (2022)
9. Bertolini, L., Gastaldi, M., Pedeferri, M.P., Redaelli, E.: Prevention of steel corrosion in concrete exposed to seawater with submerged sacrificial anodes. Corros. Sci. **44**, 1497–1513 (2002)
10. Biondini, F., Bontempi, F., Frangopol, D.M., Malerba, P.G.: Probabilistic service life assessment and maintenance planning of concrete structures. J. Struct. Eng. ASCE **132**(5), 810–825 (2006)
11. Boubitsas, D., Tang, L.P., Utgenannt, P.: Chloride ingress in concrete exposed to marine environment – field data up to 20 years' exposure. SBUF report 12684, Final draft report: 2014-02-14, CBI Betonginstitutet, Stockholm, Sweden (2014)
12. Bouteiller, V., Tissier, Y., Marie-Victoire, E., Chaussadent, T., Joiret, S.: The application of electrochemical chloride extraction to reinforced concrete - a review. Constr. Build. Mater. **351**, 128931 (2022)
13. Cady, P.D.: Inflation and highway economic analysis. J. Transp. Eng. **109**(5), 631–639 (1983)
14. CIS: Code of practice for concrete construction for water transportation projects (JTJ268-96). China Industrial Standards, Beijing, China (1996)
15. CIS: Technical specification for corrosion prevention of concrete structures in harbor engineering (JTJ 275-2000). China Industrial Standards, Beijing, China (2000)
16. CNBS: Determination of carbonation depth in hardened concrete (NC 355: 2004). Cuba National Bureau of Standards, Cuba (2004)
17. CNBS: Hardened concrete – determination of the water absorption by capillarity (NC 345: 2011). Cuba National Bureau of Standards, Cuba (2011)
18. CNBS: Hydraulic concrete – specifications (NC 120: 2014). Cuba National Bureau of Standards, Cuba (2014)
19. CNS: Technical Standard for Inspection of Building Structures (GB/T 50344–2004), China National Standard. China Architecture Publishing, Beijing (2004)
20. Dhandapani, Y., Joseph, S., Bishnoi, S., Kunther, W., Kanavaris, F., Kim, T., Irassar, E., Castel, A., Zunino, F., Machner, A., Talakokula, V., Thienel, K.C., Wilson, W., Elsen, J., Martirena, F., Santhanam, M.: Durability performance of binary and ternary blended cementitious systems with calcined clay: a RILEM TC 282-CCL review. Mater. Struct. **55**, 145 (2022)
21. Feliú, S., González, J.A., Feliú, S., Jr., Andrade, C.: Confinement of the electrical signal of in-situ measurement of polarization resistance in reinforced concrete. ACI Mater. J. **87**(5), 457–460 (1990)
22. Fjendbo, S.: Prediction of chloride ingress profiles in concrete by machine learning. In: Ilki, Çavunt, Çavunt (eds.) Proceedings of *FIB* Symposium on Building for the future: Durable, Sustainable, Resilient, Istanbul, Turkey, 5–7 June 2023. Springer (2023)
23. Fjendbo, S., Sørensen, H.E., De Weerdt, K., Geiker, M.R.: The square root method for chloride ingress prediction – applicability and limitations. Mater. Struct. **54**, 61 (2021)
24. Frederiksen, J.M., Mejlbro, L., Nilsson, L.-O.: Fick's 2nd law – complete solutions for chloride ingress into concrete. Report TVBM-3146, Lund Institute of Technology, Lund, Sweden (2008)
25. Frederiksen, J.M., Nilsson, L.-O., Sandberg, P., Tang, L.P., Andersen, A.: HETEK. A system for estimation of chloride ingress into concrete. Theoretical background. Report No. 83, The Danish Road Directorate, Denmark (1997)
26. Frederiksen, J.M., Nilsson, L.-O.: Concrete studies. Femernbelt fixed link. Chloride induced corrosion. ALECTIA, report, 2009-12-15 (Rev. 01 – 2015-03-20), Lund, Sweden (2015)
27. HZMBA: Bidding documents for general contracting of design and construction of HZM project main works in artificial islands and tunnel. HZM Bridge Administration, Zhuhai, China (2010) (in Chinese).
28. Li, C., Ng, M.K., Tang, Y.Z., Fung, T.: From a 'World Factory' to China's bay area: a review of the outline of the development plan for the Guangdong-Hong Kong-Macao Greater bay area. Plan. Theory Pract. **23**(2), 310–314 (2022)
29. Li, K.F., Li, Q.W., Zhou, X.G., Fan, Z.H.: Durability design of the Hong Kong–Zhuhai–Macau sea-link project: principle and procedure. J. Bridge Eng. ASCE **20**(11), 04015001 (2015)

30. Li, K.F., Wang, P.P., Li, Q.W., Fan, Z.H.: Durability assessment of concrete structures in HZM sea link project for service life of 120 years. Mater. Struct. **49**, 3785–3800 (2016)
31. Li, K.F., Zeng, J.J., Tang, L.P., Sorensen, H.E., Castro, B.P., et al.: Long-term field exposure of structural concretes in marine environment: state-of-the-art review by TC 289-DCM. Mater. Struct. **55**, 205 (2022)
32. Li, K.F., Zhang, D.D., Li, Q.W., Fan, Z.H.: Durability for concrete structures in marine environments of HZM project: Design, assessment and beyond. Cem. Concr. Res. **115**, 545–558 (2019)
33. Li, Q.W., Li, K.F., Zhou, X.G., Zhang, Q.M., Fan, Z.H.: Model-based durability design of concrete structures in Hong Kong–Zhuhai–Macau sea link project. Struct. Saf. **53**, 1–12 (2015)
34. Lichtblau, S.M.: Treatment of recycled aggregate for the production of durable structural concrete. Ph.D. thesis, Universidad Central "Marta Abreu" de las Villas, Santa Clara, Cuba (2019)
35. Lo, Y., Lee, H.M.: Curing effects on carbonation of concrete using a phenolphthalein indicator and Fourier-transform infrared spectroscopy. Build. Environ. **37**(5), 507–514 (2002)
36. Martirena, F., Díaz, E., Rocha, D., Maraghechi, H., Scrivener, K.: Performance of concrete made with a calcined clay-limestone-Portland cement exposed to natural conditions. In: Pam, B. (ed.) 6th International Conference on Durability of Concrete Structures, Leeds, West Yorkshire, UK, 18–20 July 2018. Whittles Publishing (2018)
37. Martirena, F., et al.: Evaluation of low carbon cement (LC3) produced concrete according to the CN 120:2014 Standards, Standardization technical committee No. 37, Ministry of Constructions, Cuba (2023)
38. Mejlbro, L.: The complete solution of Fick's second law of diffusion with time-dependent diffusion coefficient and surface concentration. In: Proceedings of Durability of Concrete in Saline Environment, Cementa AB, Danderyd, Sweden, pp. 127–158 (1996)
39. Nordtest: Concrete, Mortar and cement-based repair materials: chloride migration coefficient from non-steady migration experiments (NT Build 492), Nordtest Method, Nordtest, Espoo, Finland (1999)
40. Poulsen, S.L., Sørensen, H.E.: Chloride ingress in old Danish bridges. In: Proceedings of 2nd International Conference on Durability of Concrete (2nd ICDC), 4–6 December 2014, New Delhi, India (2014)
41. Poulsen, S.L., Sørensen, H.E., Jönsson, U.: Chloride ingress in concrete blocks at the Rødbyhavn marine exposure site – status after 5 years. In: Proceedings of 4th International Conference on Service Life Design for Infrastructures (SLD4), 27–30 August 2018, Delft, The Netherlands (2018)
42. Rebolledo, N., Andrade C.: From corrosion rate to accumulated corrosion depth or loss in cross section of reinforcements. In: Andrade, C., Gulikers, J. (eds.) Proceedings of the 4th International PhD RILEM Workshop "Advances in modelling Concrete Service Life", 19 November 2010, Madrid, Spain, Springer, pp. 43–52 (2010)
43. Selander, A.: Hydrophobic impregnation of concrete structures-effects on concrete properties. Ph.D. thesis, Royal Institute of Technology, Sweden (2010)
44. Sharma, M., Bishnoi, S., Martirena, F., Scrivener, K.: Limestone calcined clay cement and concrete: a state-of-the-art review. Cem. Concr. Res. **149**, 106564 (2021)
45. Sørensen, H.E., Jönsson, U., Mathiesen D., Munch-Petersen, C.: Femern model for chloride ingress into concrete – 2021 update. In: Proceedings of International IABSE Conference on Assessment, Upgrading and Refurbishment of Infrastructures, IABSE Rotterdam Congress Report May 2013, Rotterdam, The Netherlands (2013)
46. Sørensen, H.E., Nilsson, L.-O.: Monitoring of reinforcement corrosion in Field exposed concrete blocks. Report, 2023-02-28. Danish Technological Institute, Taastrup, Denmark (2023)
47. Tang, L.P.: Chloride ingress in concrete exposed to marine environment – field data up to 10 years' exposure. SP report 2003:16. SP Swedish National Testing and Research Institute, Borås, Sweden (2003)

48. Torrent, R.J.: A two-chamber vacuum cell for measuring the coefficient of permeability to air of the concrete cover on site. Mater. Struct. **25**, 358–365 (1992)
49. Vizcaíno-Andrés, L., Sánchez-Berriel, S., Damas-Carrera, S., Pérez-Hernández, A., Scrivener, K., Martirena, F.: Industrial trial to produce a low clinker, low carbon cement. Mater. Constr. **65**, 1–11 (2015)
50. Wang, S.N., Su, Q.K., Li, K.F.: Durability Design and Construction Technology for the Concrete Structures of Hong Kong-Zhuhai-Macau Bridge. China Communication Press, Beijing (in Chinese) (2018)
51. Yang, L.H., Li, K.F., Pang, X.Y.: Design and optimization of maintenance strategies for a long life-span port project. Mater. Struct. **46**, 161–172 (2013)

# Chapter 7
# General Conclusions and Perspectives

**Kefei Li, Junjie Zeng, Luping Tang, Carmen Andrade, Riccardo Maddalena, Mette Rica Geiker, Henrik Erndahl Sørensen, Javier Sanchez Montero, and Xuemei Liu**

## 7.1 Conclusions

1. The use of marine field exposure sites for testing and investigating structural concrete materials needs careful planning, persistence in observation, onerous onsite labor and heavy laboratory works. Nevertheless, field exposure stations provide a unique data source for the response of concrete materials to the actual

K. Li (✉)
Civil Engineering Department, Tsinghua University, Beijing 100084, China
e-mail: likefei@tsinghua.edu.cn

J. Zeng
CCCC Fourth Harbor Engineering Co., Ltd., Guangzhou 510290, China

L. Tang
Department of Architecture and Civil Engineering, Chalmers University of Technology, 41296 Gothenburg, Sweden

C. Andrade
International Center for Numerical Methods in Engineering (CIMNE), 08034 Barcelona, Spain

R. Maddalena
School of Engineering, Cardiff University, Cardiff CF24 3AA, UK

M. R. Geiker
Department of Structural Engineering, Norwegian University of Science and Technology, 7491 Trondheim, Norway

H. E. Sørensen
Building and Construction, Danish Technological Institute, Taastrup, Denmark

J. S. Montero
Eduardo Torroja Institute for Construction Sciences (IETcc), 28033 Madrid, Spain

X. Liu
Department of Infrastructure Engineering, The University of Melbourne, Parkville, Australia

 
K. Li et al. (eds.), *Long-Term Performance of Structural Concretes in Marine Exposure Conditions*, RILEM State-of-the-Art Reports Volume,
https://doi.org/10.1007/978-3-032-06959-7_7

environmental actions. These data support the understanding of the underlying mechanisms of deterioration and the mastering of durability performance to ensure the required service life of concrete infrastructures and structures. Through the comprehensive works engaged during the life of RILEM TC 289-DCM, the whole picture of marine exposure practice was constituted by the characterization of exposure sites, the definition of exposure relevant information, the identification of scientific and engineering models to help the interpretation, the establishment of a technical guideline for designing and operating exposure sites, and the support to service life management for concrete structures in severe marine environments.

2. The concrete specimens are affected by multiple deterioration mechanisms under the environmental actions in marine exposure. One distinct feature is that on the concrete surface more than one processes are active simultaneously which makes the chloride transport near concrete surface difficult to predict. In the ATM zone, the surface is subject to chloride deposition and carbonation; in the SPL zone the concrete surface is exposed to strong moisture transport and the chloride accumulation; in the TID and SUB zones the leaching will dominate the concrete surface, together with erosion and probably ice abrasion. The marine environment affects the durability of concrete structures via environmental, material and service state factors. A qualitative correlation between these factors and the durability performance was established. Among the environmental factors, the drying-wetting action is in many cases the primary impact factor on the long-term performance of structural concrete and increases the corrosion risk of embedded steel bars. For material factors, a dense pore structure and concrete surface quality have most important impact on the durability performance. For service factors, the cracking, instead of the straightforward loading level, is responsible for the altered durability performance, and cracking with allowed opening width has minor impact on the corrosion of embedded steel bars.
3. The technical characteristics of marine exposure sites, relevant to exposure practice, include the geographical location and the orientation of exposure sites, the functional planning of exposure area, the meteorology of surrounding atmosphere, the atmospheric corrosivity, the hydrology of seawater, the biological growth rates (if any), and the data collecting and recording. All environmental factors have notable impact on the durability processes. The binder type and the water to cement (binder) ratio are significant factors for long-term durability and the protections, such as silane impregnation treatment, on specimens are also of significance. The inherent dispersion of exposure data can be quantified through the tools and models from engineering statistics. With basic probability models for discrete and continuous random variables, the model parameters can be generated from observed values and the accuracy can be improved with number of measurements. The sampling procedure and the test methods on exposure specimens are crucial for collecting reliable exposure data, the operation of drilling the powder samples and the necessary amount of powder sample are two important technical aspects for sampling, and the extraction operation is important for testing of chloride contents. Finally, the exposure data are normalized both

to complete the essential information on exposure practice to support better the interpretation of exposure data, and to facilitate the data comparison and exchange for the sake of valorizing the long-term exposure practice.

4. Reliable engineering models are needed to interpret the obtained long-term exposure data and to predict the future durability performance. During the TC works, a systematic and comprehensive review has been conducted for scientific and engineering models available in literature and engineering design practice. The review began with widely-used models based on Fick's 2nd law, followed by more complex models accounting for other physical and chemical processes than chloride diffusion. Benchmarking exercises provide a structured comparison of the models, offering insights into their accuracy, applicability, and limitations. The review of benchmarking results confirmed that large dispersion exists in prediction of chloride ingress, especially for long-term durability performance. Though simple and robust models are preferred by engineering, more and more evidence shows that single diffusion-based model cannot give accurate prediction for complex environmental actions. This situation calls for the development of advanced mechanism-based models on the one hand, and robust durability indicators for engineering use on the other hand. TC works attempt to propose new indicators for engineering use including the resistivity, the tortuosity and the total chloride ingress. Enriched with more data, these indicators are expected to reinforce the available durability indicators for service life prediction.
5. The reflection and the synchronized opinions on exposure practice finally led to a technical guideline for the design and operation of long-term exposure sites in marine environments. The guideline covers the design of exposure site, the monitoring of environmental conditions, the monitoring of exposed specimens, the data management, and the maintenance of exposure facilities. The design of exposure site includes the choice of exposure sites, the exposure infrastructure and facilities, the exposure specimens and their placing and protection. The environmental factors to be monitored and the corresponding monitoring methods are then defined. For the exposed specimens are specified the target performance to monitor, the sampling, the testing and the reporting of exposure results. The data management deals with such aspects as data storage, data transmission and data sharing. The principles for the maintenance of the exposure facilities are given in the end. This guideline summarizes the best practice of long-term exposure in marine environments, expected to promote the exposure practice and to generate added values for engineering. Through several major engineering applications, one can see the substantial, even indispensable, role of exposure practice for engineering in marine environments.

## 7.2 Perspectives

The mastering of exposure practice in marine environments is far from accomplished even after the dedicated efforts of TC works, and persistent efforts should, at least, be aimed at the following issues.

1. One of the interesting topics which has not been investigated sufficiently is a similitude study of different exposure sites. The idea behind is rather straightforward: if the resemblance of durability performance can be confirmed and even quantified, the exposure data from one specific exposure site can help to predict the performance in another exposure sites, which will valorize greatly the exposure data. Certainly, this similitude should be investigated in terms of the main environmental factors and the active mechanisms. This constitutes an appealing future research topic.
2. The analysis and quantification of the inherent dispersion of data in exposure practice need to be investigated further. The TC works lay a preliminary basis for this topic, focusing on the durability parameters. Actually, the appropriate probabilistic models and the involved statistical variables should be studied for materials parameters, environmental actions and the dispersion associated with the exposure operations such as placing and sampling. With these established models, the prediction errors can be handled in a more scientific way, and applying the probabilistic models inversely can provide rational estimate for required numbers and sizes of concrete specimens for given errors. The foregoing works on the mechanical properties and structural reliability theory can provide solid support in this aspect.
3. As for the data collecting of marine exposure practice, an everlasting topic is to increase the efficiency of sampling and testing, which embraces multiple scientific questions and technical needs. It is of vital importance to improve the testing techniques to obtain maximum information from a minimum amount of collected samples since exposed specimens will be never sufficient for long observation duration. In other terms, improving the testing techniques can use less amount of sampling, and the same number of specimens can support longer observation. Besides, more portable and non-destructive devices, with the associate automatic data transmission, will greatly reduce the labor investment.

MIX
Papier aus verantwortungsvollen Quellen
Paper from responsible sources
FSC® C105338

If you have any concerns about our products,
you can contact us on
**ProductSafety@springernature.com**

In case Publisher is established outside the EU,
the EU authorized representative is:
**Springer Nature Customer Service Center GmbH**
**Europaplatz 3, 69115 Heidelberg, Germany**

Printed by Libri Plureos GmbH
in Hamburg, Germany